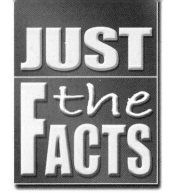

THE HUMAN BODY

By

Steve Parker

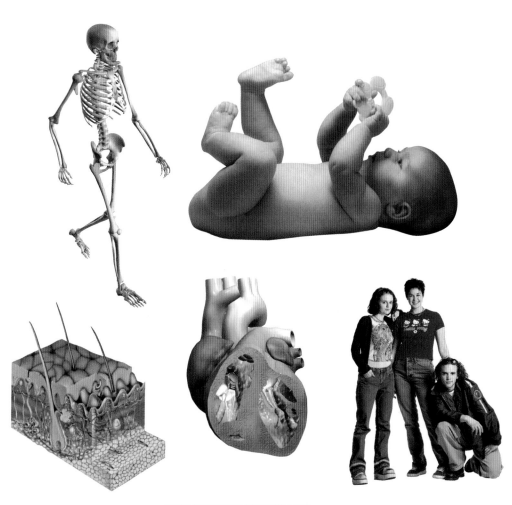

ticktock

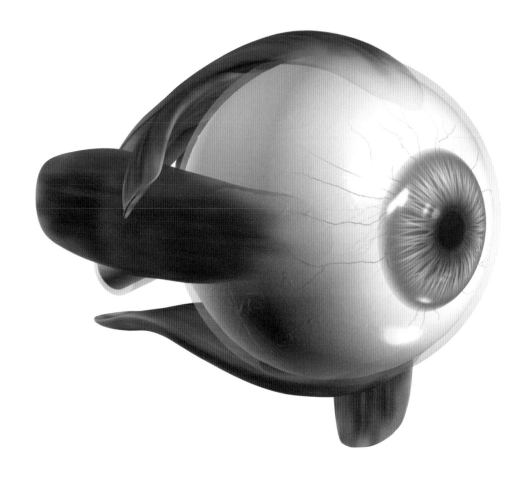

This Edition Published 2007 for Index Books Ltd
First published in Great Britain in 2006 by ticktock Media Ltd.,
Unit 2, Orchard Business Centre, North Farm Road, Tunbridge Wells, Kent, TN2 3XF
ISBN 1 86007 858 3 pbk
Printed in China
A CIP catalogue record for this book is available from the British Library.

Picture credits (t=top, b=bottom, c=centre, r=right, l=left)
Alamy: 20l, 26c, 29c, 41b, 51bl. Mediscan: 30b. PhotoAlto: 10cl. Pictor: 46cl. Primal Pictures12cr, 12b, 14t, 14c, 16tl,
16bl, 17br, 19tl, 19tr, 23t, 25t, 28c, 28b, 31br, 32br, 33, 34b, 43tc, 43tr, 43br, 44c, 52t, 53tl, 53t, 54bl. Science Photo
Library: 7 (hormonal system), 9t, 13t, 16tr, 18tl, 21t, 26b, 29b, 32t, 35, 39t, 42r, 44b,45t, 46b, 49t, 49bl, 50
(embryo br), 51 (fetus br), 53br. Tony Stone: 10b, 13b, 18bl, 19b, 34c, 36b, 37b, 39b, 45b, 51 (new baby). US Fish
and Wildlife Service: 14b, 53tcl. Wellcome Photo Library: 50 (all from week one).

CONTENTS

HOW TO USE THIS BOOK

JUST THE FACTS, THE HUMAN BODY is an easy-to-use, quick way to look up facts about the systems that control how our bodies work. Every page is packed with cut-away diagrams, charts, scientific terms and key pieces of information. For fast access to *just the facts*, follow the tips on these pages.

WHERE IN THE BODY?
An at-a-glance look at where the part of the body can be found.

INTRODUCTION TO TOPIC

TWO QUICK WAYS TO FIND A FACT:

1 Look at the detailed **CONTENTS** list on page 3 to find your topic of interest.

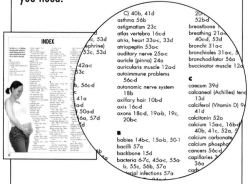

Turn to the relevant page and use the **BOX HEADINGS** to find the information box you need.

2 Turn to the **INDEX** which starts on page 60 and search for key words relating to your research.
• The index will direct you to the correct page, and where on the page to find the fact you need.

WHERE IN THE BODY?

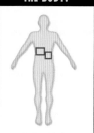

The liver is in the upper abdomen, behind the lower right ribs. The pancreas is in the upper left abdomen, behind the stomach.

LIVER & PANCREAS

Your body can't digest food with just its digestive tract (passageway) – mouth, gullet, stomach and intestines. Also needed are two parts called the liver and pancreas. These are next to the stomach and they are digestive glands, which means they make powerful substances to break down food in the intestines. Together with the digestive tract, the liver and pancreas make up the whole digestive system.

WARM LIVER

The liver is so busy with chemical processes and tasks that it makes lots of heat.

• When the body is at rest and the muscles are still, the liver makes

up to one-fifth of the body's total warmth.

• The heat from the liver isn't wasted. The blood spreads out the heat all around the body.

See pages 34-35 for information on the circulatory system.

GALL BLADDER AND BILE

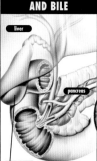

The gall bladder is a small storage bag under the liver.

• It is 8 cm long and 3 cm wide.

• Some of the bile fluid made in the liver is stored in the gall bladder.

• The gall bladder can hold up to 50 millilitres of bile.

• After a meal, bile pours from the liver along the main bile duct (tube), and from the gall bladder along the cystic duct, into the small intestine.

• Bile helps to break apart or digest the fats and oils in foods.

• The liver makes up to one litre of bile each day.

THE LIVER'S TASKS

The liver has more than 500 known tasks in the body – and probably more than haven't yet been discovered. Some of the main ones are:

• Breaking down nutrients and other substances from digestion, brought direct to the liver from the small intestine.

• Storing vitamins for times when they may be lacking in food.

• Making bile, a digestive juice.

• Breaking apart old, dead, worn-out red blood cells.

• Breaking down toxins or possibly harmful substances, like alcohol and poisons.

Helping to control the amount of water in blood and body tissues.

Alcohol is a toxin which the liver breaks down and makes harmless. Too much alcohol can overload the liver and cause a serious disease called cirrhosis.

• If levels of blood sugar (glucose) are too high, hormones from the pancreas tell the liver to change the glucose into glycogen and store it.

• If levels of blood sugar (glucose) are too low, hormones from the pancreas tell the liver to release the glycogen it has stored.

42

BOX HEADINGS
Look for heading words linked to your research to guide you to the right fact box.

JUST THE FACTS
Each topic box presents the facts you need in short, quick-to-read bullet points.

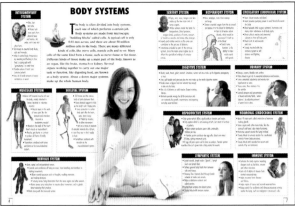

6-7 Body Systems

58-59 Glossary

HOW THE PANCREAS WORKS

Fatty foods, such as chips, are broken apart by enzymes made in the pancreas.

- Pancreas has two main jobs.
- One is to make hormones.
- The other is to make digestive chemicals called pancreatic juices.
- These juices contain about 15 powerful enzymes that break apart many substances in foods, including proteins, carbohydrates and fats.
- Pancreas makes about 1.5 litres of digestive juices daily.
- During a meal these pass along the pancreatic duct tubes into the small intestine, to attack and digest foods there.

• See page 52 for information on hormones.

WHEN THINGS GO WRONG

A yellowish tinge to the skin and eyes is known as jaundice, and it is often a sign of liver trouble.

Usually the liver breaks down old red blood cells and gets rid of the colouring substance in bile fluid. If something goes wrong the colouring substance builds up in blood and skin and causes jaundice.
Hepatitis, an infection of the liver, can cause jaundice.

UNUSUAL SUPPLY

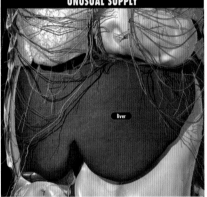

liver

One of the liver's main functions is to break down nutrients for the body. This means the liver has a unique blood supply.

- Most body parts are supplied with blood flowing along one or a few main arteries.
- The liver has a main artery, the hepatic artery.
- The liver also has a second and much greater blood supply.
- This comes along a vessel called the hepatic portal vein.
- The hepatic portal vein is the only main vein that does not take blood straight back to the heart.
- It runs from the intestines to the liver, bringing blood full of nutrients from digestion.

• See pages 36-37 for information on the blood.

BABY LIVER

Most babies and young children have big tummies (abdomens). This is partly because their liver is much larger, in proportion to the body's overall size, than the liver of an adult.

- An adult liver is usually ¼₀th of total body weight.
- A baby's liver is nearer ½₀th of total body weight.

By the time a baby becomes a toddler, their liver isn't such a large proportion of their total body weight.

WHAT IS THE LIVER?

liver

The liver is the largest single part or organ inside the body.

- Wedge-shaped, dark red in colour.
- Typical weight 1.5 kg.
- Depth at widest part on right side 15 cm.
- Has a larger right lobe and smaller left lobe.
- Lobes separated by a strong layer, the falciform ligament.

WHAT IS THE PANCREAS?

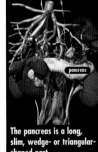

pancreas

The pancreas is a long, slim, wedge- or triangular-shaped part.

- It is soft, greyish-pink in colour.
- Typical weight 0.1 kg.
- Typical length 15 cm.
- Has three main parts: head (wide end), body (middle) and tail (tapering end).

43

EXTRA INFORMATION

The black box on the right hand side of the page explains a new aspect of the main topic.

LINKS

Look for the purple links throughout the book. Each link gives details of other pages where related or additional facts can be found.

• See pages 36-37 for information on the blood.

CUTAWAY DIAGRAMS

Clear, accurate diagrams show how the parts of the body fit together.

GLOSSARY

• A GLOSSARY of words and terms used in this book begins on page 58.

• The glossary words provide additional information to supplement the facts on the main pages.

PICTURE CAPTIONS

Captions explain what is in the pictures.

BODY SYSTEMS

- Skin, hair and nails.
- Protect soft inner parts from physical wear and knocks, dirt, water, Sun's rays and other harm.
- Skin keeps in essential body fluids, salts and minerals.
- Helps to control body temperature by sweating and flushing to lose heat, or going pale with 'goosebumps' to retain heat.
- Provides sense of touch (see Sensory system).
- Gets rid of small amounts of waste substances, in sweat.

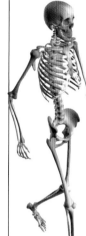

The body is often divided into body systems, each one of which performs a certain job. Body systems are made from microscopic 'building blocks' called cells. A typical cell is only 0.03 mm across, and there are about 50 million million cells in the body. There are many different kinds of cells, like nerve cells, muscle cells and so on. Many cells of the same kind form a tissue, such as nerve tissue or fat tissue. Different kinds of tissue make up a main part of the body, known as an organ, like the brain, stomach or kidney. Several organs working together to carry out one major task or function, like digesting food, are known as a body system. About a dozen major systems make up the whole human body.

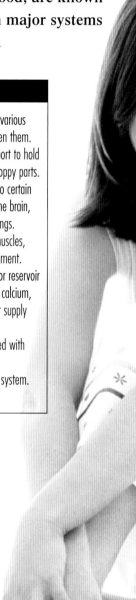

MUSCULAR SYSTEM

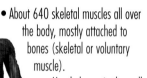

- About 640 skeletal muscles all over the body, mostly attached to bones (skeletal or voluntary muscle).
- Muscle layers in the walls of inner parts like the stomach and intestines (visceral or involuntary muscle).
- Muscle in the walls of the heart (heart muscle or myocardium).
- Muscles get shorter or contract to produce all forms of bodily movement.
- Sometimes combined with bones and joints as the musculoskeletal system.

SKELETAL SYSTEM

- 208 bones and the various kinds of joints between them.
- Gives physical support to hold up the body's soft, floppy parts.
- Gives protection to certain body parts like the brain, eyes, heart, lungs.
- Pulled by muscles, to allow movement.
- Acts as a store or reservoir of valuable minerals like calcium, in case these are in short supply in food.
- Sometimes combined with muscles as the musculoskeletal system.

NERVOUS SYSTEM

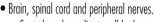

- Brain, spinal cord and peripheral nerves.
- Controls and coordinates all body processes, from breathing and heartbeat to making movements.
- Allows mental processes such as thoughts, recalling memories and making decisions.
- Sensory nerves bring information from the sense organs and other sensors.
- Motor nerves carry instructions to muscles about movement, and to glands about releasing their products.
- Works along with the hormonal system.

SENSORY SYSTEM

- Eyes, ears, nose, tongue and skin make up the five main sets of sense organs.
- Also sensors inside the body for temperature, blood pressure, oxygen levels, positions of joints, amount of stretch in muscles and many other changes.
- Gravity and motion sensors in the inner ear contribute to the process of balance.
- Sometimes included as part of the nervous system, since the main sense organs are in effect the specialized endings of sensory nerves.

RESPIRATORY SYSTEM

- Nose, windpipe, main chest airways and lungs.
- Obtains essential oxygen from the air around, and passes it to the blood for distribution.
- Gets rid of waste carbon dioxide, which would be poisonous if it built up in the blood.
- Useful 'extra function' is the ability to make vocal sounds and speech.

CIRCULATORY (CARDIOVASCULAR) SYSTEM

- Heart, blood vessels and blood.
- Heart provides pumping power to send blood all around the body.
- Blood spreads vital oxygen, nutrients, hormones and many other substances to all body parts.
- Blood collects wastes and unwanted substances from all body parts.
- Blood clots to seal wounds and cuts.
- Closely involved with the immune system in self-defence and fighting disease.

DIGESTIVE SYSTEM

- Mouth, teeth, throat, gullet, stomach, intestines, rectum and anus make up the digestive passageway or tract.
- Liver, gall bladder and pancreas plus the tract make up the whole digestive system.
- Breaks down or digests food into nutrients tiny enough to take into the body.
- Gets rid of leftovers as solid wastes (bowel motions, faeces).
- Nutrients provide energy for all life processes and raw materials for growth, maintenance and repairing everyday wear-and-tear.

URINARY SYSTEM

- Kidneys, ureters, bladder and urethra.
- Filters blood to get rid of unwanted substances and wastes.
- Forms unwanted substances and wastes into liquid waste or urine.
- Stores urine, then releases it to the outside.
- Controls amount and concentration of blood and body fluids, 'water balance', by adjusting amount of water lost in urine.

REPRODUCTIVE SYSTEM

- Only system which differs significantly in females and males.
- Only system which is not working at birth, but starts to function at puberty.
- Male system produces sperm cells continually, millions per day.
- Female system produces ripe egg cells, about one every 28 days, during menstrual cycle.
- If egg cell joins sperm cell to form an embryo, female system nourishes this as it grows into a baby inside the womb.

HORMONAL (ENDOCRINE) SYSTEM

- About 10 main parts called endocrine or hormone-making glands.
- Some organs with other main tasks, like the stomach and heart, also make hormones.
- Hormones spread around the body in blood.
- Closely linked to nervous system for coordinated control of inner body processes.
- Closely linked with reproductive system and controls it by sex hormones.

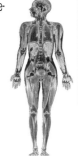

LYMPHATIC SYSTEM

- Lymph vessels, lymph nodes ('glands'), lymph ducts and lymph fluid.
- Gathers general body fluids from between cells and tissues.
- One-way flow channels fluid through lymph network of nodes and vessels.
- Helps to distribute nutrients and collect wastes.
- Lymph fluid empties into blood system.
- Closely linked with immune system.

IMMUNE SYSTEM

- Defends the body against invading dangers such as bacteria, viruses and other microbes.
- Gets rid of debris in tissues from normal wear-and-tear.
- Helps recovery from disease and illness.
- Helps repair of injury and normal wear-and-tear.
- Keeps watch for problems and disease processes arising inside the body, such as malignant (cancerous) cells.

THE SKIN

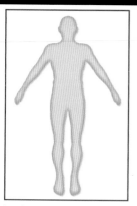

The skin is a tough but flexible layer that covers the entire body. It helps to control temperature and protects internal organs from damage.

When you look at yourself in the mirror, most of what you see is dead! Your skin, hair and nails are not living. But just underneath this dead surface, skin is very much alive, and very busy too – as you see and feel if you're unlucky enough to scratch or cut yourself. Skin is the body's largest single main part, or organ. It has at least ten main tasks, which include providing your sense of touch. It wears away every month – but it replaces itself every month, too.

SKIN MICROPARTS

An average patch of skin 1 sq cm (the size of a fingernail) contains:

- 5 million microscopic cells of at least 12 main kinds

- 100 tiny holes or pores for releasing sweat

- 1,000 micro-sensors of about six main shapes, for detecting various features of touch.

- 100-plus hairs.

- About 1 m of tiny blood vessels.

- About 50 cm of micro-nerves.

- About 100 of the tiny glands that make sebum, a natural waxy-oily substance that keeps skin supple and fairly waterproof.

TOUCH

Your sense of touch or feeling is more complicated than it seems. It's not just a single sense, detecting physical contact. It's a 'multi-sense' detecting:

- Light contact, such as a brush from a feather.

- Heavy pressure, such as being pushed or squeezed hard.

- Cold, like an ice-cube.

- Heat, such as a hot shower or bath.

- Movement, including tiny fast to-and-fro vibrations - your fingertip skin can detect vibrations which are too small for your eyes to see.

- Surface texture, such as rough wood or smooth plastic.

- Moisture content, from dry sand to wet mud.

Wearing warm clothes in winter helps protect our skin from feeling the cold.

DANGEROUS SWEAT

A person can lose 5-8 litres of sweat before the body suffers from the loss of important salts and minerals.

EXTRA SENSITIVE

- Skin on the fingertips has more than 3,000 micro-sensors per sq cm, to give the most sensitive touch.

- It has more sweat glands, to make a thin layer or film of sweat on the skin that helps you to grip better.

- It also has tiny ridges or swirls to give even better grip. These form the pattern of your fingerprints.

- Every set of fingerprints for every person around the world is different — even between identical twins.

A thin layer of sweat on the fingertips helps you to grip better. Try washing your hands thoroughly, drying them well, then picking up a paperclip.

SWEAT FACTS

Total number of sweat glands	**3-5 million**
Total length of tubes in all sweat glands stretched out straight and joined end to end	**50 km**
Amount of sweat on average day	**0.3-0.5 litres**
Amount of sweat even on a cold day	**0.07 litres**

LAYERS OF THE SKIN

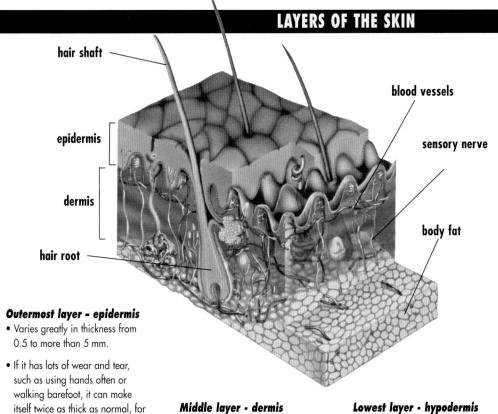

hair shaft

epidermis

dermis

hair root

blood vessels

sensory nerve

body fat

Outermost layer - epidermis
- Varies greatly in thickness from 0.5 to more than 5 mm.

- If it has lots of wear and tear, such as using hands often or walking barefoot, it can make itself twice as thick as normal, for extra protection.

Microscopic cells at its base multiply fast, fill with the tough substance keratin, move outwards, become flatter and die, and form the surface layer which is continually rubbed away.

Middle layer - dermis
- Contains sweat glands, hair roots (follicles), most of the micro-sensors for touch and tiny blood vessels called capillaries.

- Also contains fibres of the substances collagen for strength and elastin for stretchiness.

Lowest layer - hypodermis
- Contains mainly body fat, which works as a cushion against knocks and pressure.

- Works as an insulator to keep in body warmth.

• See pages 36-37 for information on blood circulation.

MICROSENSORS

- The largest touch micro-sensors are called Pacinian sensors. They have many layers like tiny onions and are up to 0.5 mm across. They detect hard pressure.

- The smallest micro-sensors are 100 times tinier and feel light touch.

SHED SKIN

- Each minute about 50,000 tiny flakes of skin are rubbed off or fall from the body.

- This loss is natural and is made up by microscopic cells at the base of the epidermis multiplying rapidly.

- This happens so fast that the epidermis replaces itself about every month.

- Over a lifetime the body sheds more than 40 kg of skin — enough to fill two typical rubbish bags.

SKIN THICKNESS

Skin makes itself thicker where it is worn or rubbed more.
On average:

• See page 23 for information on eyes.

- Soles of feet 5 mm or more
- Back 3-4 mm
- Palms of hands 2-3 mm
- Scalp on head 1.5 mm
- Fingertips 1 mm
- Average over body 1-2 mm
- Eyelids 0.5 mm

MAIN TASKS OF THE SKIN

Protection
- The skin provides protection from knocks and bumps

- It keeps out dirt, germs and liquids like water

- Shielding the body from the Sun's dangerous rays (especially UV, ultra-violet), perhaps by going darker (suntan)

Keeping fluids in
- Keeping in valuable body fluids, minerals and salts

Touch
- Providing sense of touch

Temperature control
- Cooling the body if it gets too hot

- Keeping heat inside the body in cold conditions

Vitamin D
- Production of an important nutrient, vitamin D, which keeps you healthy

Waste removal
- Removal of some waste products (in sweat)

Anti-germ layer
- Production of germ-killing substances to form a layer on skin

SIZE OF THE SKIN

Area
A typical adult's skin, taken off and ironed flat, would cover some 2 sq m – about the area of a single bed or a small shower curtain.

Weight
The weight of the skin is about 3-4 kg for a typical adult – twice as heavy as the next-largest organ, the liver.

HAIR & NAILS

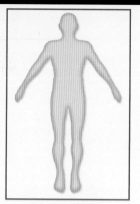

Hair is found almost all over our bodies. Nails grow at the end of each toe and finger.

H airs and nails, like the outer layer of skin (epidermis), are dead. Your body has hairs all over apart from a few places, like your palms and the palm sides of your fingers, and the soles of your feet. However some hairs grow thicker and longer, and so we notice them more. These are the hairs on the head (scalp), eyebrows and eyelashes. As we grow up, hairs also appear under the arms (axillary hair) and between the legs (pubic hair).

WHY HAVE EYELASHES AND EYEBROWS?

Eyelashes and eyebrows draw attention to our eyes. They also perform some useful functions.

EYEBROW HAIRS
Help to stop sweat dripping into the eyes.

Eyelash hairs
Help to whisk away bits of windblown dust, dirt and pests like insects.

NAIL GROWTH

- Most nails grow about 0.5 mm each week.
- In general, fingernails grow faster than toenails.
- Nails grow faster in summer than in winter.
- If you're right-handed, nails on your right hand grow faster than those on your left.
- And the other way round if you're left-handed.

EYEBROWS AND EYELASHES

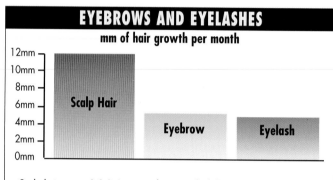

mm of hair growth per month

- Scalp hairs grow 0.3-0.4 mm each day, which is about 10-12 mm each month.
- Eyebrow hairs grow slowly, only 0.15 mm per day, usually reaching a greatest length of 10 mm.
- Eyelash hairs grow at a similar rate to eyebrow hairs, but usually stop growing at 7-8 mm long.

- See pages 22-23 for information on eyes.

THE THICKNESS OF A HAIR

- Most scalp hairs are around 0.05 mm thick, so 20 in a row would measure 1 mm.
- Fair or blonde hairs are usually thinner than dark or black scalp hairs.
- Eyelashes are thicker, up to 0.08 mm.

NAIL PARTS

There are many technical words to describe the finger nails.

Nail plate
The main flat part of the nail.

Free edge
The end of the nail which you trim, not attached to the underlying finger or toe.

Nail bed
The underside of the nail plate, which is attached to the underlying flesh but slides slowly along as it grows.

Lunula
The pale 'half-moon' where the youngest part of the nail emerges from the flesh of the finger or toe.

Eponychium
The cuticle fold where the nail base disappears under the flesh of the finger or toe.

Nail root
The growing part of the nail, hidden in the flesh of the finger or toe.

HAIR STRUCTURE AND THICKNESS

- Hairs are glued-together rods of dead, flattened, microscopic cells filled with the tough, hard body substance called keratin.

- A hair grows at its root, which is buried in a pocket-like pit in the skin called the follicle.

- Extra cells are added to the root, which pushes the rest of the hair up out of the skin.

- The part of the hair above the root is called the shaft.

• See pages 8-9 for information on the skin.

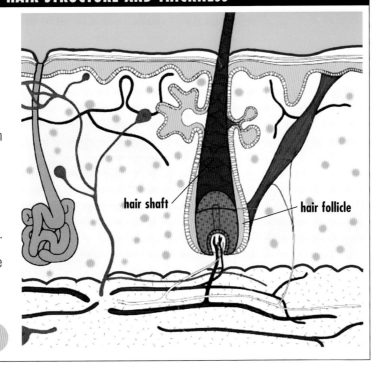

hair shaft

hair follicle

HAIR GROWTH

Different kinds of hairs grow at different rates.

- Because most scalp hairs grow for only 3-5 years, their maximum length is 50-80 cm before they fall out and are replaced.

- However some people have unusual hair that falls out much less often, and can reach lengths of 7-8 m.

Hair growth in thin, fair hair is slower than in thick, dark hair.

NUMBER OF HAIRS

The number of hairs on the head varies according to the colour of the hair. In a typical adult, the number is:

Fair hair	130,000
Brown	110,000
Black	100,000
Red	90,000

• See page 51 for information on signs of ageing.

HAIR LIFE CYCLES

Most kinds of hairs grow for a time, gradually slow down in growth rate, then 'rest' and hardly grow at all.

- After this 'rest' they usually fall out and are replaced by new hairs growing up from the same follicles (pits) in the skin.

- This means, on average, about 100 hairs are lost from the head every day.

- In eyebrow hairs the life cycle lasts about 20 weeks.

- In eyelash hairs it lasts around 10 weeks.

- In scalp hairs this life cycle lasts up to 5 years.

FASTER HAIR GROWING

- Hair growth is faster at night than during the day.

- Hair growth is faster in summer than in winter.

- Hair growth is faster around the ages of 15-25 years than before or after.

H A I R
WHY HAVE IT?

Hairs were probably more effective at their jobs millions of years ago, when they were longer, and when we looked more like our closest cousins, the apes – chimps and gorillas.

Protection
- Scalp hair protects against bangs and bruises.

- It also shields the top of the head, and the delicate brain inside, from the Sun's fierce heat or icy snowy winds.

Warmth
- Body hairs stand on end when you're cold, each pulled by a tiny muscle attached to its root, called the erector pili muscle.

- This 'hair-raising blanket' around the body helps to trap air and so keep in body warmth.

Safety
- Our hair can also stand on end when we feel frightened. When our body hairs were longer, in prehistoric times, the 'hair-raising' also made us look bigger and more impressive to enemies.

N A I L S
WHY HAVE THEM?

A nail is a strong, stiff, dead, flat plate made of the same dead substance as hairs, keratin. Each nail acts as a flat, rigid pad on the back of the fingertip.

Touch
- When you press gently on an object, the fingertip is squeezed between it and the nail.

- This makes it easier to judge pressure and the hardness of the object. Without a nail, the whole fingertip would bend back.

Scratching
- You also use nails to scratch, get rid of objects on the skin – and maybe pick your nose!

MUSCLES & MOVEMENT

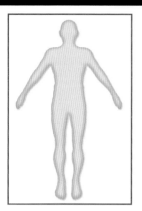

Every move you make, every breath you take, every song you sing... muscles power all of your body's movements, from blinking to leaping high in the air. Muscle actions are controlled by messages travelling to them from your brain, along nerve called motor nerves. Muscles are the body's biggest single system – a group of main parts that all work together to carry out one or a few vital tasks.

• See pages 20-21 for information on the brain.

SAVE ENERGY – GIVE A SMILE

- All muscles need energy to work, which is brought by the blood in the form of blood sugar (glucose).

- You use about 40 facial muscles to frown, but only half as many to smile. So save energy by smiling more!

• See pages 36-37 for information on blood.

TYPES OF MUSCLE

The body has three main kinds of muscles: skeletal, visceral and cardiac.

- Skeletal muscles are mostly attached to the bones of the skeleton and pull on them to make you move.

- These are the ones we normally mean when we talk about 'muscles'.

- Skeletal muscles are also called voluntary muscles because you can control them at will, voluntarily when you wish, by thinking.

- Skeletal muscles are also called striped or striated muscles because under the microscope they have a pattern of stripes or bands.

- Visceral muscles form sheets, layers or tubes in the walls of the inner body parts (viscera) like the stomach, guts and bladder.

- Visceral muscles are also called involuntary muscles because you cannot control them – they work automatically.

- Visceral muscles are also called smooth muscles because under the microscope they lack any pattern of stripes or bands.

- The third type of muscle is cardiac muscle, which forms the walls of the heart.

• See pages 32-33 for information on the heart.

Skeletal muscles, seen from the back.

MUSCLES THAT MAKE FACES

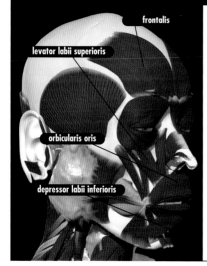

frontalis

levator labii superioris

orbicularis oris

depressor labii inferioris

We use our muscles to communicate and send information – and not just by speaking, which uses around 40 muscles. We also use muscles to 'make faces' or facial expressions. Here are some of the 60 or so face muscles and what they do.

Muscle name	Site	What it does	Expression
Frontalis	Forehead	Raises eyebrows	Surprise
Procerus	Between eyes	Pulls eyebrows in and down	Stern, concentration
Auricularis	Above and to side of ear	Wiggles ear (only in some people!)	
Buccinator	Cheek	Moves cheek	Blowing, sucking
Risorius	Side of mouth	Pulls corner of mouth	Grin
Depressor labii	Under lip	Pulls lower lip down	Frown

Face muscles allow us to make a huge range of expressions.

INSIDE A MUSCLE

A muscle is a bundle of fibres. These bundles are called fascicles. Within each fibre is a group of fibrils. A single fibril contains myosin and actin filaments. These slide past each other to shorten the muscle.

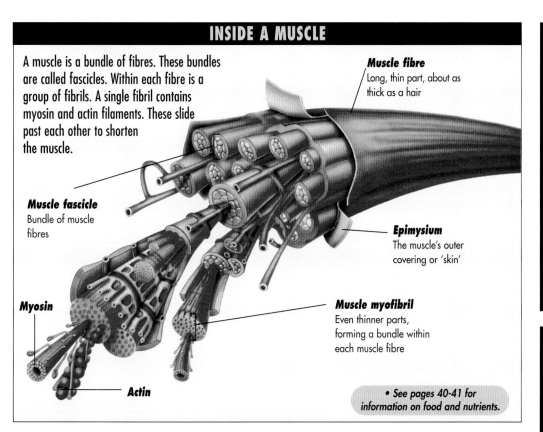

Muscle fibre
Long, thin part, about as thick as a hair

Muscle fascicle
Bundle of muscle fibres

Epimysium
The muscle's outer covering or 'skin'

Myosin

Muscle myofibril
Even thinner parts, forming a bundle within each muscle fibre

Actin

• See pages 40-41 for information on food and nutrients.

MUSCLE POWER COMPARED

This list shows the power of the body's muscle compared to various machines, in watts (the scientific units of power).

Laser-pen pointer	0.002
Heart by itself	2
All the body's muscles working hard	100
Family car on the motorway	100,000
Space shuttle	10,000 million

INDIVIDUAL VARIATIONS

• Some people have very small versions of certain muscles or none at all, as part of the natural variation between different people.

• For example, a few people lack the thin, sheet-like muscle in the neck, called the platysma.

HOW MUSCLES WORK

Muscles work by getting shorter or contracting, and pulling their ends closer together.

• In most skeletal muscles the ends taper to rope-like tendons, which are joined firmly to bones.

• Muscles cannot push or forcefully get longer, they are stretched longer when other muscles work elsewhere.

• Muscles contain two body substances or proteins called actin and myosin, which are shaped like long threads.

• In each muscle millions of these threads slide past each other to make the whole muscle shorten.

• Most muscles can shorten or contract to about two-thirds their resting length.

• A muscle bulges in the middle when it shortens but its overall size or volume does not change.

Exercise can increase the size of muscles but they have no effect on the actual number of muscles or the number of muscle cells – this stays the same.

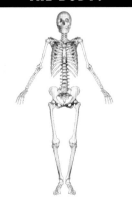

THE SKELETON

Your skeleton consists of all the bones in your body – over 200 of them. It's like an inner framework that supports the softer, floppier body parts such as the guts, nerves and blood vessels. Your skeleton is not fixed and stiff. It is a moving framework that muscles pull into hundreds of different positions every day.

BONES NOT JOINED TO OTHER BONES

There are three bones in the body not joined to any other bone.

Hyoid
A U-shaped bone in the front of the upper neck, near the throat and the base of the tongue.

Kneecap or patella
This is inside a muscle tendon and slides over the front of the knee joint, helping to protect it.

SIZE AND VARIATIONS

Our bones are a strong inner framework that hold up the soft inner parts of the body.

- There is no truth in the old belief that men and women have different numbers of ribs. Both have 24, as 12 pairs.

- However the total number of bones varies slightly as part of natural differences between people.

- For example, about one person in 20 (man or woman) has an extra pair of ribs, making 21 pairs instead of the usual 20.

- There are more bones, over 300, in the skeleton of a baby.

- As the baby grows, some of these enlarge and join or fuse together to make bigger single bones.

- The skeleton forms about one-seventh of the body's total weight.

• See pages 50-51 for information on stages of life.

TAIL END

The lowest part of the backbone is called the coccyx.
It's made of three to five smaller bones joined or fused together into one, and shaped like a small prong. It is all that's left of the long tail that our very distant ancestors had, millions of years ago, when they looked like monkeys and lived in trees.

Monkeys and humans are descended from the same distant ancestors.

WHAT ARE BONES LIKE?

Imagining our bodies as various everyday objects can help us to understand how they work.

Levers
The long bones of the arms and legs work like levers, with their pivot or fulcrum at the joint.

A bicycle chain
The many separate bones or vertebrae of the backbone only move slightly compared to each other. But over the whole backbone, this movement adds up to allow bending double, like the links of a chain such as a bicycle chain.

A cage
The ribs work like the moveable bars of a cage. This protects the heart and lungs, yet gets bigger and smaller as the lungs breathe in and out.

An eggshell
The dome shape of the cranium around the brain is a very strong design, like an eggshell. Any sharp ridges or corners would weaken it.

• See pages 16-17 for information on bones and joints.

NOT ALL BONE

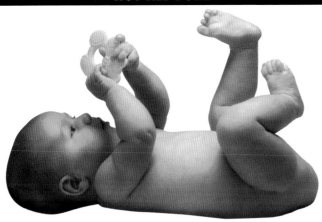

Most of a baby's skeleton is made of cartilage, not bone.

- Most bones of the skeleton begin not as real bone, but as a slightly softer, bendier, smooth substance called cartilage (gristle).

- In a developing baby, the shapes of the eventual bones form first as cartilage.

- Then as the baby grows into a child, the cartilage shapes become hardened into real bones.

- Even in the adult skeleton, some bones stay as partly cartilage.

- For example, the front end of each rib, where it joins to the breastbone, is made not of bone but of cartilage called costal cartilage.

- The nose and ears are mainly cartilage, not true bone, which is why they are slightly bendy.

SKELETON STRENGTH

Our skeleton is made of living bones that can mend themselves if broken.

- The bones of the skeleton are stronger, size for weight, than almost any kind of wood or plastic.

- If the skeleton was made of steel, it would weigh four times as much.

- The thigh bone can stand a pressure of 3 tonnes per sq cm when we leap and land.

- The skeleton can also mend itself, which no kind of plastic or metal can.

NUMBERS OF BONES

A human skeleton contains, on average, 206 bones. They are divided up through the body as follows:

Skull

Cranium (brain case) 8

Face 14

Ear 3 tiny bones each

Total: 28 bones

Throat (hyoid bone) 1

Backbone

Neck (cervical vertebrae) 7

Chest (thoracic vertebrae) 12

Lower back (lumbar vertebrae) 5

Base of back (sacrum, coccyx) 2

Total: 26 bones

Rib cage

Ribs 24

Breastbone 1

Total: 25 bones

Arms

Shoulder 2

Upper arm 1

Forearm 2

Wrist 8

Palm 5

Fingers and thumb 14

Total: 32 bones in each arm (includes hand)

Legs

Hip 1

Thigh and knee 2

Shin 2

Ankle 7

Sole of foot 5

Toes 14

Total: 31 bones in each leg (includes foot)

SKELETON'S MAIN TASKS

The main tasks of the skeleton are to:

- Hold up the body, giving support to softer parts.

- Allow the body to move, when pulled by muscles.

- Provide openings for the nose and mouth, for breathing and eating.

- Protect certain body parts, for example, the upper skull around the brain, the front skull around most of the eyes, and the ribs around the lungs and chest.

- Store many body minerals such as calcium and magnesium, for times when food is scarce and these minerals are in short supply for other body processes like sending nerve messages.

- Make new microscopic cells for the blood, at the rate of 3 million every second. These cells are produced in the soft jelly-like bone marrow found in the centres of some bones.

The skeleton provides some protection for vital body parts, but it is wise to use a helmet when you are doing sport that involves going fast.

BONES & JOINTS

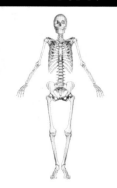

Joints allow the skeleton
to move. They can be found
all over the body.

Your skeleton of bones holds you up. But you would not be able to move if it wasn't for the joints which link your bones together. More than half of your body's bones – 112 out of 206 – are in your wrists, hands, fingers, ankles, feet and toes. So are more than half of your 200-plus joints. Your bones, muscles and joints work so closely together that they are sometimes viewed as a single system, called the musculo-skeletal system.

PARTS OF A BONE

Periosteum
The outer covering or 'skin' wrapped all around the bone.

Foramen
Small hole in a bone, where a nerve or blood vessel passes to its inside.

Compact bone
Very strong, hard outer layer of bone, like a shell.

Osteons (Haversian systems)
Tiny 'rods' of bone substance bundled and glued together to make compact bone.

Spongy or cancellous bone
Inner layer of a bone, under the compact bone, which has holes like a sponge.

Marrow
Jelly-like substance in the centre of most bones.

Red marrow
Found in all bones of a baby, but in an adult, only in the long bones of the arms and legs, ribs, backbone, breastbone and upper skull. Makes new microscopic cells for the blood.

Yellow marrow
In adults, found mainly in smaller bones of the hands and feet. Contains fat for use as an energy reserve, but can change to red marrow if needed, for example, after serious bleeding.

BONE MAKE UP

- The name 'skeleton' comes from an ancient word meaning 'dried up'. But living bones are not dry, they are about one-quarter water. (Overall, the body is two-thirds water.)

- The main minerals in bone are calcium, phosphate and carbonate.

These form tiny crystals which give bone its hardness and stiffness.

- Bone also contains tiny thread-like fibres of the substance collagen, which makes it slightly bendy under pressure, so it is less likely to snap.

- If a bone is soaked in a special acid chemical, this gets rid of all the crystals of calcium phosphate and calcium carbonate. It leaves only the collagen fibres, which are so bendy that a long bone like the upper-arm bone can be tied in a knot!

During long space flights, the lack of gravity means that bones are put under little pressure. They start to lose minerals and become weaker. Astronauts exercise regularly to keep their bones strong.

- See pages 40-41 for information on food and nutrients.

LIGAMENTS

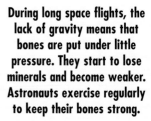

Ligament   Muscle

Bones are held together at a joint by stretchy straps called ligaments, which stop them moving too far or coming apart. If the bones slip and come out of their usual position, this is called a dislocation.

YES AND NO

The two topmost backbones (cervical vertebrae) of the spine have special joint designs. They allow the head to make important movements.

- The atlas (uppermost) is more like a ring. It supports and allows the head to turn or rotate to look to the side as when saying 'no'.

- The axis (second uppermost) has a curved shape like a saddle. It allows the head to tilt to the side and nod as when saying 'yes'.

PARTS OF A BONE

Doctors have names for each part of a bone.

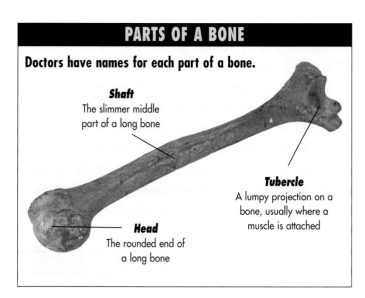

Shaft
The slimmer middle part of a long bone

Tubercle
A lumpy projection on a bone, usually where a muscle is attached

Head
The rounded end of a long bone

BONE RECORDS

Longest
Thigh bone (femur) forming about one-quarter of total body height.

Widest
Hip bone (pelvis), forming the body's broadest part.

Smallest
Stirrup (stapes) deep inside the ear, a U-shape just 8 mm long.

Toughest
Lower jaw (mandible), used hundreds of times daily when chewing.

DESIGN OF THE JOINT

The different designs of your body's joints are sometimes compared to machines and mechanical gadgets.

Hinge joint
Allows the bones to move only to and fro, not side to side (as in a door hinge).

Examples: knee, smaller knuckles of fingers.

Ball-and-socket joint
Allows the bones to move to and fro and also side to side, and perhaps twist (rotate).

Examples: hips, shoulders, larger knuckles.

Saddle joints
Shaped like a saddle for tilting and sliding.

Example: thumb.

Washer joints
Limited tilting with a pad or washer of cartilage between the bone ends.

Examples: joints between the backbones, where the cartilage pad is called the disc (intervertebral disc).

Fixed or suture joint
No movement at all, because the bones are firmly joined together.

Examples: between the bones of the cranium (upper skull) around the brain.

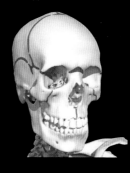

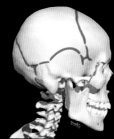

There are eight bones in the cranium. They are fused together to protect the brain underneath.

REDUCING WEAR AND TEAR

- Where the ends of a bone touch in a joint, they are covered with smooth, glossy cartilage, to reduce wear and rubbing.

- The space between the bones is filled with a slippery liquid called synovial fluid, which works like oil in a machine to reduce wear even more.

- The fluid is kept in by a loose bag around the joint: the joint capsule.

- New synovial fluid is always being made by the inner lining of this bag, called the synovial membrane.

Even a large joint like the hip contains only about a teaspoon of synovial fluid.

BIGGEST JOINT

Your single biggest joint, your knee, has an unusual design with extra cartilages and ligaments.

- As well as cartilage covering the ends of the thigh and shin bones, it has two pieces of curved, moon-shaped cartilage between these bones.

- The cartilage pieces are called menisci and help the knee to 'lock' straight so you can stand up easily.

- When sports people have 'torn knee cartilage' it's usually one of these menisci which is damaged.

• See pages 12-13 for information on muscles.

- The knee has two strong sets of ligaments, the lateral ligaments on the outer side and the medial ligaments on the inner side (next to the other knee).

- As well as these, it has two ligaments inside, keeping the ends of the bones very close together.

- These two ligaments form an X-like cross shape and are called cruciate ligaments.

Playing sport can sometimes damage your knee joint. It is important to always warm up before playing sport.

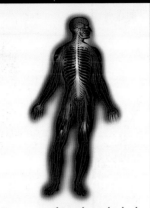

Nerves run throughout the body, carrying electrical signals from the brain.

THE NERVOUS SYSTEM

You are nervous, I am nervous, everyone is nervous. That is, we all have a nerve system to control our every movement and action, and every process that happens inside the body. Your nervous system is made up of your brain, spinal cord and nerves. It works by sending tiny electrical signals called nerve impulses. Millions of these travel around the body and brain every second, like the busiest computer network.

MAIN PARTS OF THE NERVOUS SYSTEM

There are two main nervous systems within the body. The central nervous system is the brain's main control centre. It sends nerve impulses to the rest of the body using the peripheral nervous system. We have conscious control over the central and peripheral nervous systems.

Central nervous system:
Brain
Inside the top half of the head.
Spinal cord
The main nerve link between the brain and the body.

Peripheral nervous system:
Cranial nerves
Connect directly to the brain rather than the spinal cord. They go mainly to parts in the head like the eyes, ears and nose (see below).
Spinal nerves
Branch out from the spinal cord to the arms, legs, back, chest and all other body parts.

• See pages 20-21 for information on the brain.

SLOW TO HURT

When you hurt a finger you probably feel the touch first, and then the pain starts a moment later. This is because the signals about touch travel faster along the nerves than the signals about pain.

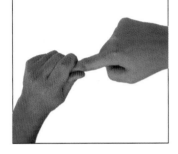

AUTOMATIC SYSTEM

• Some parts of the nerve system work automatically, without you having to think about them.

• They are called the autonomic nerve system.

• They control inner processes like heartbeat, digesting food, body temperature and blood pressure.

We have no conscious control over some parts of our body, such as the systems that control digestion.

NERVES AND NERVE CELLS

A nerve's outer covering is called the epineurium. Inside are bundles of nerve fibres or axons, each too small to see without a microscope.

epineurium

axons

• See pages 8-9 for information on the skin.

NERVE SIGNALS

A nerve signal is a tiny pulse or peak of electricity, made by moving chemical substances into and out of the nerve cell.

Average signal strength	$\frac{1}{10}$th of a volt.
Average signal length	$\frac{1}{1000}$th of a second.
Average recovery time before another signal can pass	$\frac{1}{500}$th of a second.
Slowest signals travel at	0.5 metres per second.
Fastest signals go	140 metres per second.

THICKEST NERVE

The sciatic nerve, in the hip and upper thigh, is about the width of its owner's thumb. This is thicker than the spinal cord, which is usually the width of its owner's little finger.

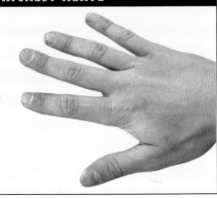

NERVES AND NERVE CELLS

Nerves are bendy but tough, so they can move easily at joints but withstand being squeezed by muscles around them.

- Each nerve fibre is the long, wire-like part of a single microscopic nerve cell, called a neuron.

- Usually near one end, the nerve cell has a wider part, the cell body.

- Branching from the nerve cell body are even thinner spidery-looking parts called dendrites.

- Nerve messages from other nerve cells are picked up by the dendrites, processed and altered as they pass around the cell body, and then sent on by the axon (fibre) to other nerve cells.

- Most nerve fibres are 0.01 mm wide, so 100 side by side would stretch one millimetre.

- They have a covering wrapped around them, called the myelin sheath. It makes nerve messages travel faster and stops them leaking away.

A typical nerve looks like shiny wire or string.

SPINAL CORD

The spinal cord, in the back, is one of the most important parts of our nervous system.

- Joins the brain to the main body.

- Is about 45 cm long in a typical adult.

- Has 31 pairs of nerves branching from it, left and right.

- Is protected inside a 'tunnel' formed by a row of holes through the backbones.

- Like the brain, has a layer of liquid around it called cerebrospinal fluid, to cushion it from knocks and sudden twists.

NERVE JUNCTIONS

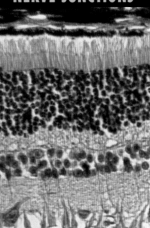

Synapses are so small that scientists have to use special electron microscopes to study them.

Each nerve cell receives signals from thousands of other cells and passes on signals to thousands more.

- Individual nerve cells do not actually touch each other where the ends of their dendrites and axons (fibres) come together.

- The ends are separated by tiny gaps, at junction points called synapses.

- The gap inside a synapse is just 0.025 micrometres wide, which means 40,000 in a row would stretch one millimetre.

- Nerve messages 'jump' across a synapse not as electrical signals, but in the form of chemicals, called neurotransmitters.

- This chemical 'jump' takes less than $\frac{1}{1000}$ th of a second.

NERVE LENGTHS

- All the nerves in the body, taken out and joined end to end, would stretch about 100 km.

- The longest single nerve fibres, in the legs, are up to one metre in length.

DIRECT TO THE BRAIN

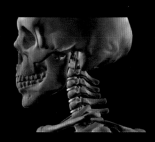

12 pairs of cranial nerves join directly to the brain, and link it to the following parts:

1. Nose
For smelling

2. Eyes
For seeing

3. Eyeball-moving muscles

4. Eyeball-moving muscles

5. Skin and touch
On forehead, face, cheeks, jaw muscles, muscles for chewing

6. Eyeball-moving muscles

7. Tongue
For taste, saliva (spit) glands, tear glands, facial expressions

8. Ear
For hearing and balance

9. Rear of tongue
For taste, swallowing muscles

10. Swallowing muscles
Also lungs and heart in chest

11. Voicebox muscles
For speaking

12. Tongue muscles
For speaking, swallowing

NERVES TO EVERY PART

There are nerves to every body part, including the heart, lungs and guts.

- The thickest ones near the brain and spinal cord are known as nerve trunks.

- The thinnest ones spreading into body parts are terminal fibres.

THE BRAIN

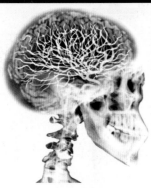

The brain is inside the cranium (the domed part of the skull), forming the upper half of the head.

Sometimes we describe a very clever person as having 'lots of brains'. But we only have one brain each. It contains more than 100 billion nerve cells, or neurons – about as many stars as in our galaxy, the Milky Way. The brain also contains perhaps ten times as many 'support' cells called neuroglia. It's not the size of a brain which makes it clever, or the exact number of cells. It depends on how often its owner uses it, and in how much detail – by looking, listening, learning, remembering, using imagination and having ideas.

HUNGRY FOR ENERGY

- The brain consumes about one-fifth of all the energy used by the body.

- But the brain forms only about $\frac{1}{50}$ th of the whole body.

- So the brain uses ten times more energy for its size, compared to most other body parts.

- This energy is mainly in the form of blood sugar or glucose, brought to the brain by its main blood vessels, the carotid and vertebral arteries.

- Average blood flow to the brain is 750 ml per minute, about one-eighth of the heart's total output.

- This flow is the same whether the body is at rest or very active.

- This is unusual because blood flow to other body parts changes greatly between rest and activity, for example, it increases to the muscle by ten times and decreases to the stomach and guts by half.

• See pages 34-35 for information on the circulatory system.

CORTEX IN CONTROL

planning movement

making movement

speech

touch on the skin

hearing

vision

- The outer grey layer of the cerebrum, over the top of the brain, is called the cerebral cortex.

- Spread out flat, it would be the area of a pillowcase – and almost as thin.

- However deep grooves called sulci mean it is wrinkled and folded into the space inside the upper skull.

- The cortex has about half the brain's total number of nerve cells, around 50 billion.

- Each of these can have connections with more than 200,000 other nerve cells.

- The connections are made by the spidery-looking 'arms' called dendrites and a much longer, wire-like part of the nerve fibre.

- The cortex is the main place where we become aware of what we see, hear, smell, taste and touch, that is, what we sense.

- It is also the place where we plan movements and actions and get them started, known as motor skills.

- Each of these sensory and motor processes takes place in a different area or patch of the cortex, known as a centre.

- The cortex is also the major site for thinking and our general awareness and consciousness – what we call our 'mind'.

- And the cortex is involved in learning and memory, although scientists aren't quite sure how.

• See pages 22-27 for information on the senses.

• See pages 12-13 for information on muscles.

THE WEIGHT OF THE BRAIN

The weight of an average adult brain is 1.4 kg.

The largest accurately measured normal human brain is 2.9 kg.

SIZE ISN'T EVERYTHING

- Bigger brains are not necessarily cleverer, and there is no link between the size of a healthy brain and intelligence.

- The average female brain is slightly smaller than the average male brain.

- But the average female body is smaller, in comparison, to the average male body.

- So compared to body size, women have slightly larger brains than men.

MAIN BRAIN PARTS

Cerebrum
The big wrinkled, domed part covering most of the top of the brain, forms more than four-fifths of the whole brain. It has a thin outer layer of 'grey matter', which is mainly nerve cells, covering an inner mass of 'white matter' which is chiefly nerve fibres.

Cerebellum
A smaller wrinkled part at the lower rear, looks like a smaller version of the whole brain. in fact its name means 'little brain'. It carries out detailed control of muscles so we can move about, keep our balance and carry out skilled actions.

Thalamus
This is two egg-shaped parts almost at the centre of the brain. It helps to sort and process information from four of the senses (eyes, ears, tongue, skin) going to the cerebrum above.

Hypothalamus
Just below and in front of the thalamus, is a main centre for powerful feelings, emotions and urges such as anger, fear, love and joy.

The brain stem
The base of the brain contains the main 'life support' areas for heartbeat, breathing, blood pressure and control of digestion. Its lower end merges into the top of the spinal cord.

HOLLOW BRAIN

- The brain has four small chambers inside it called ventricles.

- These are filled with a pale liquid called cerebrospinal fluid, CSF.

- CSF is found around the brain, between two of the protective layers called meninges which surround it. CSF is also found inside and around the spinal cord.

- The total amount of CSF inside and around the brain and spinal cord is about 125 ml.

- This fluid flows very slowly and is gradually renewed about three times every 24 hours.

- CSF is important as it helps to cushion the brain from knocks.

- The liquid also supports the brain within the skull, brings nourishment and takes away wastes.

- If someone has an epidural anaesthetic, this is injected into the meninges and CSF around the lower spinal cord.

SLEEP

Even when asleep, the brain is just as active sending nerve messages around itself as it is when awake. This is shown by recordings of its electrical nerve signals.

- Older people tend to sleep more hours overall but often in several sessions, such as 'cat-naps' through the day.

- Usual sleep needs for most people every 24 hours:

New baby	*20 hours*
10-year-old	*10 hours*
Adult	*7-8 hours*

LEFT AND RIGHT

- Nerve messages from the body cross over from left to right at the base of the brain.

- This means the left side of the brain receives signals from, and sends them to, the right side of the body.

- In most people the left side of the brain is more active in speaking and reading, scientific skills, using numbers and maths, and working out problems in a step-by-step way.

- The right side of the brain is more active in dealing with shapes and colours, artistic skills like painting and music, having ideas, 'jumping' to answers without detailed thought.

- In a right-handed person, the left side of the brain is generally dominant. In a left-handed person, the right side of the brain is generally dominant.

A person may write with one hand, but use the other to carry out everday tasks.

THE GROWING B R A I N

The development of the brain happens quickly after conception. It continues to grow in size after birth, and makes new nerve connections throughout childhood.

Inside the womb
- The brain is one of the first main body parts to form just three weeks after conception, as a large arched bulge.

Four weeks after conception
- The brain is almost larger than the entire rest of the body.

20 weeks after conception
- Brain weighs about 100 grams.

At birth
- The brain is 400-500 grams, about one-third of its final adult size. In comparison, a new baby's body is about ½5th of its final adult size.

Growing up
- By age 3 years the brain is almost fully grown and weighs 1.1 kg.

- The brain does not make any new nerve cells after birth.

- But it does make new connections between nerve cells, perhaps millions every week, as we take in knowledge, develop skills and learn new things.

From the age of about 20 years
- The brain shrinks by about one gram of weight per year. This represents the loss of around 10,000 nerve cells each day.

- Certain drugs, including alcohol, can speed this cell loss and make the brain shrink faster.

• See pages 50-51 for information on the stages of life.

EYES & SIGHT

Each eyeball is in a bony bowl called the eye socket. It is formed by curved parts of five skull bones.

For people with normal sight, sight is the most precious sense. Experts guess that over half of the information we have in our brains came in through our eyes, as words (like these), pictures, drawings, real-life scenes and images on screens. Yet the eye does not really 'see'. It turns patterns of light rays into patterns of nerve signals, which go to the brain. The visual or sight centre at the back of the brain is the 'mind's eye', where we recognise and understand what we see.

BLIND SPOT

There are no rods or cones at the place where all the retina's nerve fibres come together to form the start of the optic nerve. This place is called the optic disc.

- Since light cannot be detected here, it is known as the blind spot.
- Normally we don't notice the blind spot because our eyes continually dart and look around at different parts of a scene.
- As we do this the brain guesses and 'fills in' the missing area from what is around and what it has seen just before or after.
- The optic nerve contains one million nerve fibres – the most of any nerve carrying sense information to the brain.

FLOATERS

- Some people see spots or 'floaters' which seem to be in front of the eye.
- These are actually inside the eye, in the vitreous humour jelly that fills the inside of the eyeball.
- They are usually stray red blood cells or bits of fibres which have escaped from the retina.
- We can't look straight at them because as we move the eyeball they move too inside it.
- A few floaters are normal, but medical advice is needed if they suddenly increase in number.

WHAT AN INCREDIBLE SIGHT

As it goes dark in the evening, what we see seems to lose colour and 'grey out'. This is because the cone cells work less well and we rely on the rods.

- The eye's inner lining, the retina, is where light rays are changed to nerve signals.
- The retina has an area about the same as a larger postage stamp (wider across than high).
- It has millions of microscopic cells that make nerve signals when hit by light rays.
- 125 million are rod cells, which work well in dim light, but cannot see colours, only shades of grey.

- 7 million are cone cells, which see fine details and colours, but work only in bright light.
- Most of the cones are concentrated in a slightly bowl-shaped hollow at the back of the retina, the fovea or yellow spot.
- This is where light falls to give us the clearest, most detailed view.

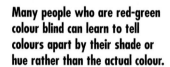
- See pages 18-19 for information on the nerves.

COLOUR CONES

Many people who are red-green colour blind can learn to tell colours apart by their shade or hue rather than the actual colour.

- There are three kinds of light-detecting cone cells in the retina.
- They are called red, green and blue cones.
- This is not because of their colours – they all look the same.
- The names are because the three types of cones detect three different colours of light.
- The brain works out the colour of an object from the active cones.

- In some people, not all these cones are present or work properly.
- This is called 'colour blindness', or more properly, defect of colour vision.
- Most common is when red and green appear similar.
- This often runs in families, and nearly always affects boys rather than girls.
- True colour blindness, seeing everything in shades of grey (like an old black-and-white movie) is very rare, affecting

MAIN PARTS OF THE EYE

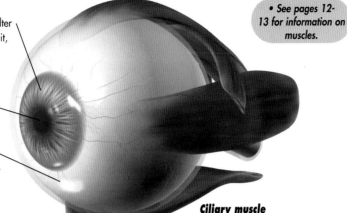

• See pages 12-13 for information on muscles.

Iris
Coloured ring of muscle which can alter the size of the hole (the pupil) within it, making it smaller in bright light to protect the delicate inside of the eye.

Pupil
Hole in the iris through which light enters the eye's interior.

Sclera
Tough outer layer or sheath around the whole eye apart from the cornea.

Cornea
Thick clear dome at the front of the eye.

Conjunctiva
Sensitive covering at the front of the eyeball, over the cornea.

Lens
Pea-shaped blob about 10 mm across which alters in shape to see and focus clearly, from looking at faraway objects to nearby ones.

Retina
Inner layer lining the eyeball's interior.

Choroid
Blood-rich layer between sclera and retina.

Ciliary muscle
Ring of muscle around the lens which alters its shape.

Aqueous humour
Thin, clear fluid filling the space between the cornea and the lens.

Vitreous humour
Thick clear jelly-like substance filling the main eyeball and giving its rounded shape.

MOVING THE EYEBALL

Behind the eyeball are six small ribbon-shaped muscles that make it turn and swivel in its socket or eye orbit.

Medial rectus
Moves the eye inwards towards the nose.

Lateral rectus
Moves the eye outwards away from the nose.

Superior rectus
Moves the eye upwards to look at the sky.

Inferior rectus
Moves the eye downwards to look at the ground.

Superior oblique
Pulls eye inwards and downwards.

Inferior oblique
Pulls eye upwards and outwards.

In total the eye can tilt as follows:

• look up by 35 degrees
• look down by 50 degrees
• inwards towards the nose by 50 degrees
• outwards by 45 degrees.

IRIS SECURITY SCANS

Usually, people with darker skin and hair have browner irises. People with lighter skin and hair have bluer irises.

• Each person has a different colour and detailed pattern of marks on the iris.

• Photos or scans of the iris fed into a computer, can be used like fingerprints for identification and security checks.

• Rarely a person has two different coloured irises, perhaps being born like that, or as a result of injury or illness.

• Every person in the world has different fingerprints, which can be used for identification and security checks.

• The same applies to the coloured part of the eye, the iris.

BLINKING AMAZING

• We spend about up to 30 minutes of our waking day with our eyes shut during blinks.

• Blinking washes soothing, cleansing tear fluid over the eye. The fluid washes away dust and helps to kill germs.

• Tear fluid comes from the lacrimal gland, just above and to the outer side of each eye, under a fold of skin.

• On average:
 Number of blinks per minute: 6
 Length of blink: 0.3-0.4 seconds
 Total amount of tear
 fluid made in a day: **50 ml**

This can treble if surroundings are dusty or have chemical fumes.

MEASURING THE EYE

The eyeball is almost a perfect sphere or ball shape. Its measurements are as follows:

Side-to-side: 24 mm.

Front-to-back: 24 mm.

Top-to-bottom: 23 mm.

• The eyeball's overall weight is 25-30 grams.

• The eye is one of the body parts that grows least from birth to adulthood.

20/20 VISION

The saying '20/20 vision' came about from the way of describing how clearly a person can see.

• 20/20 means a person can see, at a distance of 20 feet, what normal eyesight can show.

• The larger the second number, the worse the eyesight.

• Someone with 20/60 vision can see at 20 feet what a normally sighted person sees clearly at 60 feet.

• Short sightedness (myopia), is due to the eyeball being too big for the focusing power of its lens, for example, 28-29 mm across.

• Long sightedness (hypermetropia), is due to the eyeball being too small for lens to focus, for example, 20-21 mm across.

• Astigmatism is when the curve of the eyeball is not regular in all directions like a bowl, but more in one direction than the other, like a spoon.

EARS & HEARING

Ssh – can you hear that sound? Hardly anywhere is truly silent. We can usually hear some kind of sound, whether the roar of a jet plane, friends talking, or birds singing and the wind rustling grass. Much of the time we are not aware of sounds around us, because they tell us nothing new. The brain 'blocks out' frequent noise like humming machinery or distant traffic. Only when we hear something new, important or exciting, does the mind turn its attention to hearing.

The outer ear is on the side of the head, usually level with the nose. The inner ear is deep in the temporal skull bone, almost behind the eye.

OUTSIDE TO INSIDE

The ear is divided into three main sections:

Outer ear
Ear flap (pinna or auricle), ear canal

Middle ear
Eardrum, tiny ear bones, middle ear chamber

Inner ear
Cochlea, semi-circular canals and their chambers.

EAR BONES

- The body's six smallest bones, three in each middle ear.

- They were named long ago from items more common at the time, to do with horseriding and ironsmiths.

- Hammer (malleus) is attached to the eardrum.

- Anvil (incus) is the middle of the three.

- Stirrup (stapes) is attached to the oval window of the cochlea.

• See pages 14-17 for information on bones.

STEREO HEARING

- Sound travels about 330 metres per second in air.

- A sound from one side reaches the ear on that side more than 1,000th of a second before it reaches the other ear.

- The sound is louder and clearer in the nearer ear too.

- The brain can detect these differences in time, volume and clarity, and work out the direction a sound comes from.

- This is known as stereophonic hearing.

- Headphones and earphones copy these differences to give the impression of wide-apart sounds.

- Even sounds in front and behind can be told apart, whether they come from low down or high up.

- A sound from the floor directly in front causes some echoes and brings these mixed in with it.

- A sound from directly above has fewer echoes and these reach the ears after the main sounds.

• See pages 20-21 for information on the brain.

THE SENSE OF BALANCE

In space, there is no gravity to help give astronauts a sense of balance. The lack of gravity causes about a third of people to get space sick.

- Three semicircular canals are at right angles to each other.

- Each canal has a jelly-like blob at one end in its widened part or ampulla.

- Stuck in the jelly-like blob are micro-hairs from hair cells.

- As the head moves, fluid in the canal swishes to and from and moves the jelly-like blob.

- This moves the hairs of the hair cells, which make nerve signals and send them to the brain.

- In the wider parts next to the canals, the utricle and saccule chambers, are more blobs with hairs in them.

- Gravity pulls these down, bending the hairs and making the hair cells produce nerve signals.

- So the canals sense head movements while the chambers detect head position.

- But balance involves much more, including information from the eyes about what is upright and level, from the skin about whether the body is leaning, and from inside the muscles and joints about strains on them.

- The brain uses all this information to adjust muscles and keep us well-balanced.

HOW WE HEAR

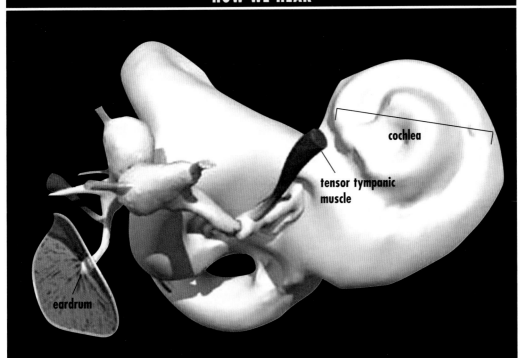

cochlea

tensor tympanic muscle

eardrum

1 Sound waves approach as invisible ripples of high and low air pressure.

2 Outer ear flap funnels sound waves.

3 Ear canal carries them into the skull.

4 Eardrum shakes fast or vibrates as sound waves bounce off it.

5 Vibrations pass along row of three tiny bones or ossicles.

6 Third ossicle makes thin 'window' of cochlea vibrate.

7 Vibrations pass into fluid inside cochlea, causing ripples.

8 Ripples shake 50-100 micro-hairs on each of 25,000 microscopic hair cells inside cochlea.

9 Hair cells make nerve signals when shaken.

10 Nerve signals pass along nerve fibres into cochlear nerve.

11 Cochlear nerve is joined by vestibular nerve from balance parts.

12 Both nerves form auditory nerve which carries nerve signals to brain.

LOUDNESS OF SOUNDS

Sound intensity (roughly the same as loudness or volume) is measured in units called decibels, dB.

0 dB Total silence

10 dB Limit of human hearing

20 dB Watch ticking

30 dB Whisper

40 dB Quiet talking, distant traffic

50 dB Normal talking

60 dB Normal television volume

70 dB Traffic in city street, vacuum cleaner

80 dB Alarm clock ringing, nearby truck

90 dB Heavy traffic at side of motor-way, music in disco or club

100 dB Chainsaw, road drill

DAMAGE CONTROL

Some sounds are too loud for us to hear comfortably. We put our hands over our ears to try to protect our ears.

- Sounds above 90 dB, especially if high-pitched like a whining or sawing, can damage hearing.

- Many places have laws controlling noise and limiting people being exposed to it, like factories, airports and music clubs.

PITCH

Sound reaches us as waves of vibrations of the air. Higher sounds make the air vibrate more quickly than lower sounds.

- Pitch is the scale of a sound — whether it makes the air vibrate at a high or low frequency.

- Our ears can detect sounds from 25 to 20,000 vibrations per second.

- Dogs can detect much deeper and much higher sounds than this.

EAR MEASUREMENTS

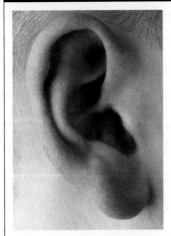

The ear canal leads from the outer ear to the eardrum. It is 20 mm long and slightly S-shaped.

Eardrum surface area 55 sq mm (about the size of the nail on the little finger).

Stirrup bone less than 5 mm long.

Cochlea spiral like a snail, with two 3/4 turns.

Cochlea 9 mm across at the wide end.

Cochlea straightened out would stretch 35 mm.

Semicircular canals each 15-20 mm long, curved into a C-shape (for balance).

Semicircular canals each less than 1 mm wide.

NOSE & TONGUE

The nasal and oral chambers – nose and mouth – form the front lower quarter of the head, each shaped by the skull and jaw bones around it.

Smell and taste are called 'chemosenses'. This means they detect chemical substances – tiny particles too small to see. The nose reacts to particles called odorants, floating in the air. The tongue does the same to particles called flavorants, in foods and drinks. Both these senses are very useful, since they can warn us of danger – but also give us plenty of pleasure.

FAST TRACK SIGNALS

Nerve signals about smell take a different route through the brain, compared to signals from other senses.

They pass through a brain part called the limbic system, which is involved in feelings and emotions. This is why a strong smell brings back powerful memories and feelings.

• See pages 20-21 for information on the brain.

INSIDE THE NOSE

Nostrils
Two holes, each leading to one side of the nasal chamber.

Nasal cartilages
Curved sheets of cartilage (gristle) forming the sticking-out part of the nose.

Nasal chamber
The air space inside the nose, roughly below the inner sides of the eyes.

Septum
Flat sheet of cartilage dividing the two halves of the nasal chamber.

Turbinates
Shelf-like ridges on each outer side of the nasal chamber.

Olfactory patch
Fuzzy-looking area inside the top of each half of the nasal chamber, which detects smells.

LOCK AND KEY

Experts are still not exactly sure how smell and taste work in detail. The main idea is the 'lock and key theory'.

• Microscopic sense cells for both smell and taste are called hair cells, with many tiny hairs sticking out, known as cilia.

• These hairs are probably coated with thousands of different-shaped receptors or 'landing pads'.

A rose produces a particular smell particle that our nose can recognise.

• Each type of smell or flavour particle has its own particular shape.

• Particles try to fit in all the receptors on the hairs, but only slot exactly into certain ones of the same shape, like a key fitting into a lock.

• When a particle fits into a receptor, the hair cell sends a nerve signal to the brain.

• The brain works out the smell or taste from the overall pattern of nerve signals it receives.

MICRO-DETAILS: THE NOSE

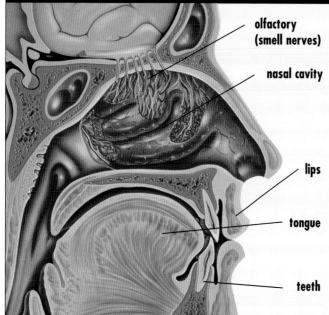

olfactory (smell nerves)

nasal cavity

lips

tongue

teeth

• The olfactory patch in the top of the nasal chamber is about the area of a thumbnail.

• Each olfactory patch has 10 million smell hair cells.

• Each smell hair cell has 10-20 micro-hairs sticking down from it.

• All the micro-hairs from one nose, joined end to end, would stretch over 100 metres.

• The micro-hairs stick into the sticky slimy mucus that lines the inside of the nasal chamber, inside the nose.

• Odorant particles floating in air seep into the slimy mucus coating the inside of the nasal chamber.

• They then come into contact with the micro-hairs.

• A single smell hair cell lives for about 30 days and is then replaced.

• Over many years some smell hair cells die but are not replaced.

• So younger people have a more sensitive sense of smell than older people.

MICRO-DETAILS: THE TONGUE

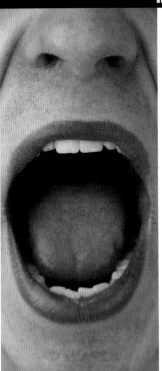

The tip, sides and rear of the tongue have about 10,000 tiny taste buds, too small to see.

- Most taste buds are around and between the little lumps or 'pimples' on the tongue, called papillae.

- Each taste bud is shaped like a tiny onion and contains about 25 taste hair cells.

- Each taste hair smell has about 10 short micro-hairs sticking up from it.

- The micro-hairs stick through a hole called a taste pore, at the top of the taste bud, onto the tongue's surface.

- Flavorant particles in foods and drinks seep into the saliva (spit)

covering the tongue and come into contact with the micro-hairs.

- A single taste hair cell lives for about 10 days and is then replaced.

- Over many years some taste hair cells die but are not replaced. So younger people have more sensitive taste than older people.

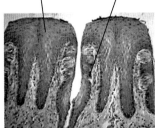

A close up view of a taste bud.

HOW MANY?

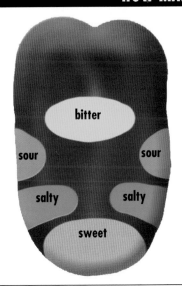

We sense different basic flavours on different parts of the tongue.

- The tongue can detect only four main flavours: sweet at the tip, salty along the front sides, sour along the rear sides, and bitter across the back.

- There are no taste buds on the main middle part of the tongue's upper surface, or below it.

NOT ALL IT SEEMS

Food is less appetising if we have a cold. The nose is full of mucus, smell does not work, and food seems less 'tasty'. In fact it's less 'smelly'.

When we 'taste' a meal, it's not only taste at work.

- Smells from food in the mouth waft up, around the back of the roof of the mouth, into the nose.

- Here they are sensed by the nose in the usual way.

- Touch sensors in the gums and

cheeks and on the tongue tell us about the food too.

- These touch sensors detect if the food is hot or cold, hard or soft, rough with bits or smooth like cream, and so on.

- So enjoying a meal involves taste, smell and touch.

NOSE AND MEMORY

With practice, most people could probably tell apart up to 10,000 different smells, odours, scents and fragrances.

However this depends on having a good memory as well as a sensitive nose.

SNIFF SNIFF

The nasal chamber inside the nose makes up to one litre of slimy mucus every day. Most we sniff in and swallow, some we blow out.

TONGUE TASKS

The tongue is the body's most flexible or bendy muscle.

It has twelve parts or sections of muscles inside it, and goes from long and thin, poking out, to short and wide at the back of the mouth, in less than a second.

In addition to taste the tongue:

Helps in eating
- Moves food around inside the mouth so it is all chewed well.
- Separates a smaller lump of food from the whole chewed mouthful, for swallowing.
- Licks bits of food off the teeth and lips.

Touches the lips
- Moistening the lips helps them seal together well.
- Stop dribbles.

Communicates
- Changes shape while speaking to make words sound clear.
- Helps to make other sounds for communication, like whistles, hisses and clicks.

TONGUE TWISTER

Usually we don't have to think about talking — the words just come out of our mouth.

When we try to say a tongue twister, we realise how difficult it can be for the brain and the tongue to work together. Try these!

- Red lorry, yellow lorry, red lorry, yellow lorry

- Which wristwatches are Swiss wristwatches?

- A skunk sat on a stump and thunk the stump stunk, but the stump thunk the skunk stunk.

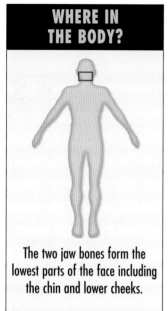

The two jaw bones form the lowest parts of the face including the chin and lower cheeks.

TEETH & JAW

Give yourself a smile in the mirror – and take a look at your teeth. Hopefully they are clean and shiny! Teeth are the hardest parts of the whole body. We use them hundreds of times each day as we bite and chew. But they are the only body parts that cannot try to mend themselves if damaged or diseased. So we must look after them well, or one day, all we'll be able to eat is soup up a straw!

WHAT TEETH DO

Bite small pieces of large items of food.

⬇

Crush food into softer pieces.

⬇

Chew these into even softer, squidgier lumps.

NUMBERS OF TEETH

Baby, milk or deciduous:
8 incisors
4 canines
8 premolars
Total: 20 in full set
Baby teeth are important because they help the adult teeth to grow into the correct shape.

Adult or permanent:
8 incisors
4 canines
8 premolars,
12 molars
Total: 32 in full set

• See page 50-51 for information on the stages of life.

TOOTH NAMES AND SHAPES

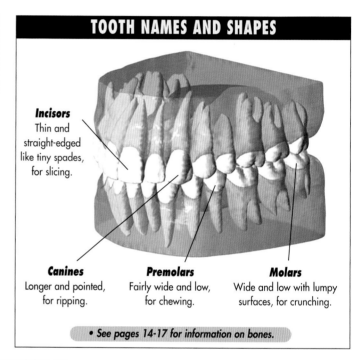

Incisors
Thin and straight-edged like tiny spades, for slicing.

Canines
Longer and pointed, for ripping.

Premolars
Fairly wide and low, for chewing.

Molars
Wide and low with lumpy surfaces, for crunching.

• See pages 14-17 for information on bones.

JAWS AND CHEWING

- The upper jaw bone is called the maxilla.

- The lower jaw bone is called the mandible.

- The mandible is the largest and strongest bone of the face.

- The mandible has some of the hardest, toughest bone in the body.

- One of the main chewing muscles is the temporalis, which runs from the temple (side of the head above the ear) to the lower side of the lower jaw.

- Another main chewing muscle is the masseter, which runs from the cheekbone to the lower side of the lower jaw.

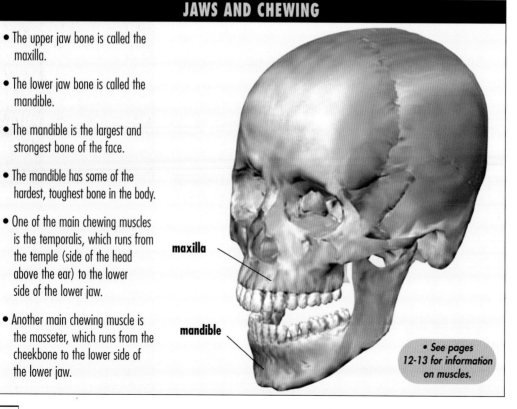

maxilla

mandible

• See pages 12-13 for information on muscles.

TWO ROOFS

The roof of the mouth has two main parts.

- The front part behind the nose is called the hard palate.

- It is formed by a backwards-facing curved plate of the upper jaw bone (maxilla) plus part of another skull bone behind this, the palatine bone.

- The rear part above the back of the mouth is the soft palate.

- This is made mainly of muscles, cartilage (gristle) and fibres.

- It can bend up as a lump of food is pushed to the back of the mouth for swallowing.

PARTS OF A TOOTH

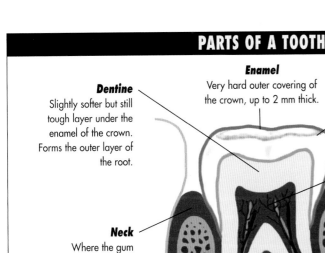

Enamel
Very hard outer covering of the crown, up to 2 mm thick.

Dentine
Slightly softer but still tough layer under the enamel of the crown. Forms the outer layer of the root.

Crown
The part of the tooth above the gum.

Dental pulp
Soft innermost layer of the tooth, made mainly of blood vessels and nerve ends.

Neck
Where the gum folds in and down around the tooth at the gum surface.

Cementum
The 'living glue' that holds the tooth's root into the jaw.

Root
The part of the tooth fixed into the jaw bone.

Root canal
Small tunnel-like hole at base of root, where the blood vessels and nerves go from the jaw bone into the pulp.

PLAQUE DANGER

- All mouths are full of bacteria, although not all are harmful.

- Without proper brushing, bacteria will form on the hard enamel of the teeth.

- The bacteria multiply and form a film over the enamel. This is called plaque.

- Sugary foods help to 'glue' the plaque on to the tooth enamel.

- Sugar will also make the plaque produce acid, which eats into the tooth enamel.

- The acid makes tiny holes in the enamel. These get bigger and are called cavities.

- The tooth does not hurt until the acid reaches the nerves. By then, the cavity is already there.

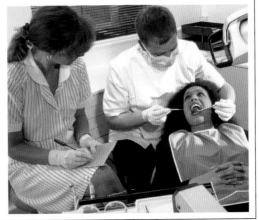

A dentist will check that your teeth are healthy and that no cavities are developing.

FALLING OUT, GROWING IN

Teeth	Time baby one appears (months)	Time adult one appears (years)
1st incisors	6-12	6-8
2nd incisors	9-15	7-9
Canines	14-20	9-12
1st premolars	15-20	10-12
2nd premolars	24-30	10-12
1st molars	–	6-7
2nd molars	–	11-13
3rd molars	–	18-21

FUNNY NAMES

- Canines are called eye teeth.

- Wisdom teeth are the rearmost molars, not appearing unto a person is grown up and supposedly more experienced and wiser than when a child.

- In some people the wisdom teeth never erupt, or grow above the gum.

FACTS ABOUT SALIVA

We could not chew and swallow without saliva or spit. It would be very difficult to eat.

Chewing
- It moistens food so it is easier to chew.

- The moist food can be squashed into a lump that slips down easily when swallowed.

- Our taste sensors work less well when food is dry, so saliva gives dry food its taste.

Enzymes
- Chemicals called enzymes in saliva begin to digest the food as it is chewed, especially starchy foods like potato, bread, rice and pasta.

Hygiene
- Saliva washes away small particles of food and helps to keep the mouth clean.

• See page 38-39 for information on digestion.

HOW SALIVA IS MADE

Saliva is made in six salivary glands around the face.

- The parotid glands are below and to the front of each ear.

- The submandibular glands are in the angle of the lower jaw.

- The sublingual glands are in the floor of the mouth below the tongue.

- Together the six glands make a total of about 1.5 litres of saliva each day.

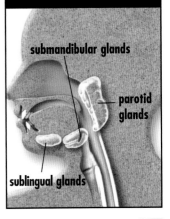

submandibular glands

parotid glands

sublingual glands

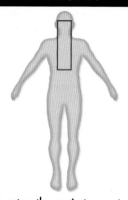

Air enters the respiratory system through the nose or mouth and travels down the windpipe to the lungs in the chest.

LUNGS & BREATHING

Puff, pant, in, out, suck, blow... The body's breathing or respiratory system obtains the vital substance oxygen from the air around us. Oxygen is needed to take part in the body chemistry that breaks apart blood sugar (glucose), releasing its energy to power almost every body process and action. The main parts of the system are the lungs, reached by the series of airways leading down through the nose, throat and windpipe.

• *See pages 26-27 for information on the nose.*

SIZE AND SHAPE OF THE LUNGS

- Each lung is shaped almost like a cone.

- The upper point or apex reaches slightly higher than the collar bone across the top of the chest to the shoulder.

- The wide base sits on the dome-shaped main breathing muscle, the diaphragm, which is roughly level with the bottom of the breastbone but curves down to the bottom ribs around the sides.

- The left lung has two main parts or lobes and a scooped-out shape where the heart fits.

- The right lung has three lobes and is on average about one-fifth bigger than the left lung.

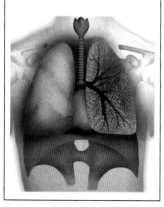

TOTAL BREATHING

- Volume of air passing through the lungs in a year: 4 million litres.

- Number of breaths in a lifetime: about 500 million.

AIR SPEEDS

Air is expelled from our lungs at different rates.

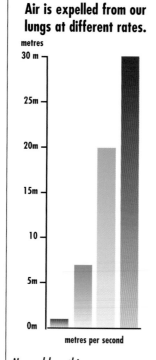

metres
30 m

25m

20m

15m

10

5m

0m

metres per second

Normal breathing –
2 metres per second.

Fast breathing –
7 metres per second.

Coughing –
20 metres per second.

Sneeze –
30 metres per second.

AIR AND BREATHING RATES

These are average volumes for an adult man. For women the amounts are about one-quarter less.

All the air in the lungs when fully breathed in:	**6 litres**
Air in the lungs left after completely breathing out:	**1.2 litres**
Air between breathing out normally, and breathing out forcefully and completely:	**1.0 litres**
Air breathed in and out at rest:	**0.5 litres**
Extra air when breathing in very forcefully:	**3.3 litres**
Normal breathing rate at rest:	**15 in-and-out per minute**
Breathing rate after great activity:	**50 per minute**
Amount of air breathed in and out after great activity:	**3 litres**

- **So the amount of air going into and coming out of lungs varies from 7.5 litres at rest to 150 litres after great activity.**

• *See page 32 for the heart rate during exercise.*

HOW AIR CHANGES

oxygen   other

nitrogen

Fresh air breathed in

79% nitrogen

20% oxygen

0.03% carbon dioxide

oxygen   other

nitrogen

Stale air breathed out

79% nitrogen

16% oxygen,

4% carbon dioxide

BRANCHING AIRWAYS

Air passes through a series of chambers and tubes on its way deep into the lungs.

The total length of all the air tubes in the lungs joined end to end — 50 km.

• See page 54 for information on the tonsils.

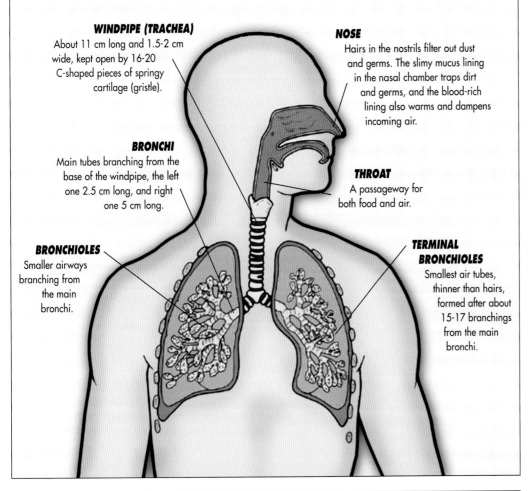

WINDPIPE (TRACHEA)
About 11 cm long and 1.5-2 cm wide, kept open by 16-20 C-shaped pieces of springy cartilage (gristle).

NOSE
Hairs in the nostrils filter out dust and germs. The slimy mucus lining in the nasal chamber traps dirt and germs, and the blood-rich lining also warms and dampens incoming air.

BRONCHI
Main tubes branching from the base of the windpipe, the left one 2.5 cm long, and right one 5 cm long.

THROAT
A passageway for both food and air.

BRONCHIOLES
Smaller airways branching from the main bronchi.

TERMINAL BRONCHIOLES
Smallest air tubes, thinner than hairs, formed after about 15-17 branchings from the main bronchi.

BREATHING AND SPEECH

Air passing out of the lungs has a useful extra effect — speech.

- There are nine pieces of cartilage (gristle) in the voice-box or larynx.

- Front ridge of the thyroid cartilage forms the 'Adam's apple' which males and females have, but which is more noticeable in males.

- About 19 muscles of the larynx alter the length of the vocal cords (vocal folds) to make the sounds of speech.

- The vocal cords are about 5 mm longer in men than women, giving them a deeper voice.

- Average pitch of male vocal cords: 120 Hz (vibrations per second).

- Average pitch of female vocal cords: 210 Hz (vibrations per second).

- Average pitch of child's vocal cords: 260 Hz (vibrations per second).

FAIR EXCHANGE

The places where oxygen is taken into the body are tiny bubble-shaped spaces deep in the lungs, called alveoli.

- Alveoli are bunched at the end of the smallest airways, the terminal bronchioles.

- There are 250-300 million alveoli in each lung.

- Breathing not only takes in oxygen, it also gets rid of the waste product carbon dioxide, which would soon poison the body if it collected.

Spread out flat, all the alveoli from both lungs would cover a tennis court.

THE VOICEBOX

Above the voicebox is the leaf-shaped flap of epiglottis cartilage. When food is swallowed this folds down over the entrance to the voicebox to prevent food entering the airway and causing choking.

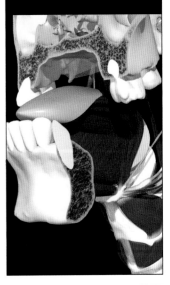

THE HEART

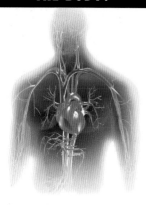

In the centre of your chest, below a thin layer of skin, muscle and bone sits your heart. This simple, yet essential, pump carries blood to and from your body's billions of cells non-stop, day and night. During an average lifetime (70 years) the heart beats 2.5 billion times. Although your heart cannot actually control whether you will fall in love or if you are a big-hearted (kind) person, without the heart's second-by-second collection and delivery service, your cells – and your body – would die.

The heart is between the lungs. It tips slightly to the left side, which is why people think it is on the left side of the body.

THE HEART'S JOB

The right and left sides of the heart work side by side like two pumps.

Every time the heart contracts, or beats, the right side pumps oxygen-poor blood back to the lungs to pick up oxygen, and the left side pumps oxygen-rich blood from the lungs out into the body.

PULSE RATE (HEARTBEATS) PER MINUTE

The number of heartbeats per minute is called the pulse rate. The resting heart rate changes throughout our lives. When we exercise, our heart needs to supply more oxygen to our bodies, so it pumps harder and faster.

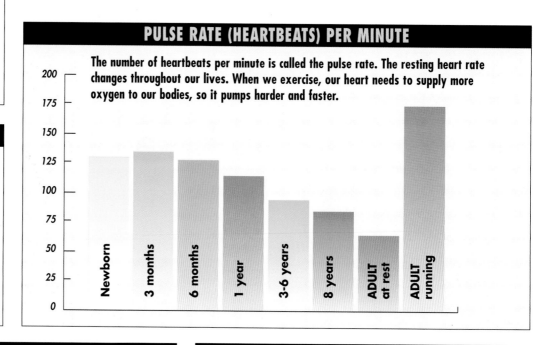

Newborn · 3 months · 6 months · 1 year · 3-6 years · 8 years · ADULT at rest · ADULT running

PHYSICAL CHARACTERISTICS

The heart is about the size of its owner's clenched fist. As you grow from a child into an adult, your heart will grow at the same rate as your clenched fist.

Average heart weight male:	**300 g**	Size: Length:	**12 cm**
Average heart weight female:	**250 g**	Width:	**8-9 cm**
		Front to back:	**6 cm**

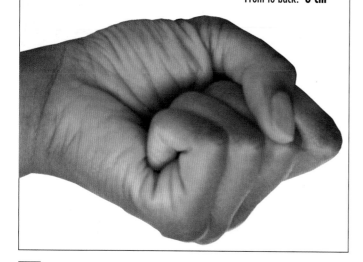

WHAT MAKES THE HEART BEAT?

At the top of the heart is a tiny area called the sino-atrial node. It sends out electrical signals that make the cells in the heart wall contract.

THE HEART'S OWN BLOOD SUPPLY

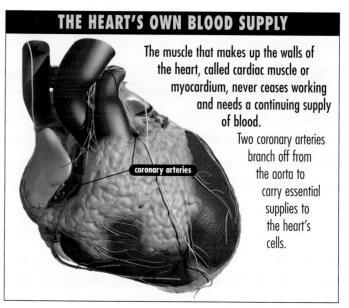

The muscle that makes up the walls of the heart, called cardiac muscle or myocardium, never ceases working and needs a continuing supply of blood.

Two coronary arteries branch off from the aorta to carry essential supplies to the heart's cells.

coronary arteries

THE HEART (VIEWED FROM THE FRONT)

Superior (upper) vena cava

Aorta

Pulmonary artery

Sino-atrial node

Right atrium

Left atrium

Pulmonary veins

Pulmonary valve

Mitral valve

Pericardium

Tricuspid valve

Left ventricle

Septum

Inferior (lower) vena cava

Right ventricle

Heartstrings

Muscular wall of cardiac (heart) muscle

PARTS OF THE
HEART

Aorta
The largest artery in the body. It is about the diameter of a garden hose.

Atrium
One of the two small upper chambers in the heart which receives blood from the veins and passes it to the ventricles below.

Heartstrings
Cords which hold the valve in place between the atrium and the ventricle.

Inferior Vena Cava
A large vein which collects blood from the lower half of the body.

Pericardium
The tough outer covering of the heart.

Septum
Muscular wall which divides the left and right sides of the heart.

Superior Vena Cava
A large vein which collects blood from the upper half of the body.

Valve
A sort of door that only opens one way to let blood through, but stops it flowing backwards. The heart has four valves: tricuspid, pulmonary, mitral, aortic.

Ventricle
One of the two large lower chambers in the heart which receives blood from the atrium above and passes it out into the arteries.

HOW THE HEART WORKS

Dotted lines represent oxygen-poor blood

•••• 1) Blood flows in from the body to the right atrium through the superior and inferior vena cavas. It is known as oxygen-poor blood because the body has taken and used the oxygen that the blood was carrying.

•••• 2) The right atrium pumps the blood through the tricuspid valve into the right ventricle.

•••• 3) The right ventricle pumps the blood through the pulmonary valve into the pulmonary artery and off into the lungs.

Inside the lungs

4) As the blood travels through the lungs, it releases waste gases and picks up oxygen.

Dashed lines represent oxygen-rich blood

‒‒‒ 5) The blood flows from the lungs into the left atrium through the pulmonary veins.

‒‒‒ 6) The left atrium pumps the blood into the left ventricle.

‒‒‒ 7) The left ventricle pumps the blood through the aortic valve into the aorta and off around the body.

CIRCULATORY SYSTEM

Something that circulates goes round and round, and that's exactly what happens to the blood inside your body. Pumped by your heart, your blood flows around a network of pipes and tubes called blood vessels, on a non-stop, never-ending journey round and round your body. The blood vessels, together with the heart itself, and also the thick red liquid called blood, all make up the circulatory system.

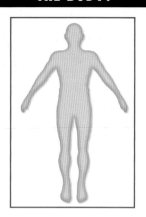

Blood vessels reach every tiny part of your body from the top of your head to the ends of your fingers and toes.

- Some parts have much fewer blood vessels than others, such as the tough, tapering tendons at the ends of muscles, which have more than 10 times fewer blood vessels than the muscle itself.

- Only a few small body parts have no blood vessels at all, for example, the lens of the eye.

DIFFERENT NAMES

The circulatory system is also called the cardiovascular system.

- 'Cardio' is to do with the heart. It comes from the ancient Greek word kardia, meaning heart.

- 'Vascular' is to do with blood vessels. It comes from the Latin word vas, meaning vessel.

- The vessels leading to and from each body part are known as its vascular supply.

MAIN VEINS AND ARTERIES

Exercise pumps the blood around the body faster.

Veins
- The main vein bringing blood from the head, arms and upper body back to the heart is called the superior vena cava.

- The main vein bringing blood from the lower body, hips and legs back the heart is called the inferior vena cava.

- Both these main veins are about 30 mm wide.

- Blood flows very slowly through them, at only 1 mm per second.

- At any single moment these main veins contain one-tenth of all the body's blood.

Arteries
- The body's main artery is the aorta, carrying blood from the left side of the heart to all body parts.

- The aorta is 40 cm long and arches up, over and down behind the heart, inside the chest.

- The aorta's width is about 25 mm.

- Its walls are 3 mm thick.

- Blood surges through the aorta at about 30 cm per second.

• See pages 36-37 for information on blood.

BLUE AND RED

Blood that has a lot of oxygen in it is red. When the body has used the oxygen the blood becomes blue.

- Arteries carry blood from the heart, but not all of this blood is 'red' or fresh and high in oxygen.

- The pulmonary arteries from the right side of the heart to the lungs, carry 'blue', stale blood low in oxygen.

- Similarly veins carry blood to the heart, but not all of this blood is 'blue' or stale and low in oxygen.

- The pulmonary veins from the lungs to the left side of the heart carry 'red' or fresh blood, high in oxygen.

• See pages 30-31 for information on the lungs.

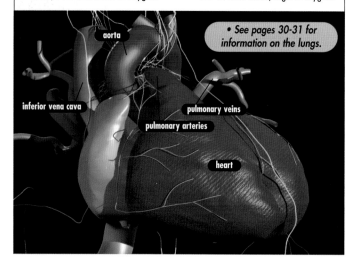

aorta

inferior vena cava

pulmonary veins

pulmonary arteries

heart

NAMING THE PARTS

Most arteries and veins are named from the body parts they supply.

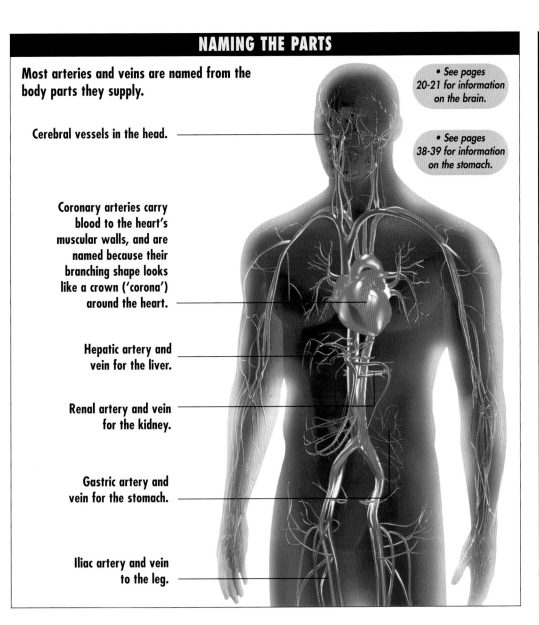

Cerebral vessels in the head.

Coronary arteries carry blood to the heart's muscular walls, and are named because their branching shape looks like a crown ('corona') around the heart.

Hepatic artery and vein for the liver.

Renal artery and vein for the kidney.

Gastric artery and vein for the stomach.

Iliac artery and vein to the leg.

• See pages 20-21 for information on the brain.

• See pages 38-39 for information on the stomach.

TYPES OF BLOOD VESSELS

Arteries
- Carry blood away from the heart.
- Thick, muscular walls to cope with the surge of high pressure as blood is forced from the heart with each heartbeat.
- Take blood to the body's main parts or organs.
- Divide and become thinner forming arterioles.

Arterioles
- Smaller and shorter than arteries.
- Muscles in the walls can tighten to make the arterioles smaller, or relax and make them wider, to control amount of blood flowing through.
- Divide and become narrower forming capillaries.

Capillaries
- The smallest blood vessels, very short and too thin to see without a microscope.
- Walls are so thin (just one cell thick) that nutrients and useful substances can pass from the blood inside, through the walls to cells and tissues around.
- Thin walls also allow wastes and unwanted substances to pass from cells and tissues around, into the blood, to be taken away.
- Join together to form venules.

Venules
- Thin-walled and very bendy.
- Collect blood from within each main body part.
- Join together forming veins.

Veins
- Wide with thin, floppy walls.
- Have pocket-like valves sticking out from their walls, to make sure blood flows the correct way.
- Carry blood from main body parts back to the heart.

BLOOD VESSEL CHART

Blood vessel	Typical diameter across (mm)	Wall thickness (mm)	Typical length (mm)	Blood pressure inside (blood emerging from heart = max 100)
Arteries	5	1	150	90
Arterioles	0.5	0.02	5	60
Capillaries	0.008	0.001	0.7	30
Venules	0.02	0.003	3	20
Veins	15	0.5	150	10

JOURNEY TIMES

The journey time from any tiny drop of blood depends on its route around the circulatory system.

- A short trip from the heart's right side to the lungs and straight back to the heart's left side can last less than 10 seconds.

- A long trip from the heart's left side all the way down through the body and legs to the toes, then all the way back the heart's right side, can last more than a minute.

• See pages 32-33 for information on the heart.

LENGTH AND AREA

- If all the body's blood vessels could be taken out and joined end to end, they would stretch about 100,000 km, which is two and a half times around the world.

- If all the capillaries were ironed flat their total surface area would be about half a football pitch.

BLOOD

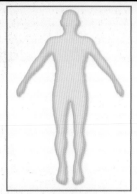

At any single moment about ⅛th of all the body's blood is in the arteries, almost three-quarters is in the veins, and less than 1/20th is in the tiny capillaries inside body parts or organs.

HOW MUCH?

The amount of blood in the body depends mainly on body size.

- On average blood forms 1/12th of the body's total weight.

- This is slightly less in women compared to men.

- Most women naturally have more fatty tissue than men, which has less blood supply compared to other body parts.

- Also most women naturally have less muscle tissue than men, which has more blood supply compared to other body parts.

- An average adult woman has 4 to 5 litres of blood.

- An average adult man has 5 to 6 litres of blood.

- For people of average weight and build, the volume of blood is about 79 millilitres per kilogram of body weight.

> • See pages 12-13 for information on muscles.

Ouch, that cut hurts! And from it oozes a thick, red liquid which every body part needs to stay alive – blood. Pumped by the heart, it flows through tubes called blood vessels. Blood carries useful substances like oxygen and nutrients to all body parts. It also collects wastes and unwanted substances, which are removed mainly by the kidneys. But apart from this delivery-and-collection service, blood does much much more...

BLOOD FLOW THROUGH BODY PARTS

In general, busier body parts need more blood supply.

- When a body part is active, changes occur in the blood vessels in order to supply it with more blood.

- The muscles in the walls of the small blood vessels called arterioles relax.

- This allows more blood to flow through them to the part they supply.

- The width of the arterioles is controlled mainly by signals from the brain sent along nerves.

- The hormone adrenaline also affects the width of the arterioles.

> • See pages 34-35 for information on the circulatory system.

BLOOD GROUPS

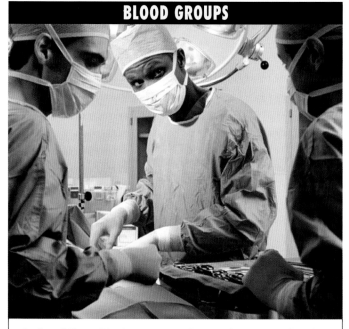

The four different blood groups were discovered in 1900. Before this, blood transfusions had a high rate of failure. Today we realise it is vital to know the blood groups of the donor and the patient in order for blood to be used safely.

- Certain kinds or groups of blood, when mixed together, may form clumps or clots.

- This can be dangerous during a blood transfusion, when blood is given or donated by one person, to be put into another person, the recipient .

- ABO is the system for testing blood for its group. A person can be either A, B, AB or O.

- A person with group O is a 'universal donor' whose blood can be given to almost anyone. A person with group AB is a 'universal recipient' who can receive blood from almost anyone.

Blood group of person	Can donate blood to	Can receive blood from
A	A, AB	A, O
B	B, AB	B, O
AB	AB	A, B, AB, O
O	A, B, AB, O	O

RED BLOOD CELLS

Red blood cells carry oxygen around the body.

- Red blood cells are among the most numerous cells, with 25,000 billion in an average person.

- They are also among the smallest cells, each one just 7 microns ($\frac{1}{140}$th of a millimetre) across and 2 microns thick.

- Each red cell is shaped like a doughnut without the hole poked completely though.

- A red cell's colour is due to the substance haemoglobin.

- Haemoglobin joins or attaches to oxygen and carries it around the body.

- Each red cell contains 250 million tiny particles, or molecules, of haemoglobin.

- Each red blood cell lives for three or four months, then dies and is broken apart.

- This means about 3 million red blood cells die every second — and the same number of new ones are made.

- Red blood cells, like white blood cells and platelets, are made in the jelly-like marrow inside bones.

Our blood contains millions and millions of red blood cells.

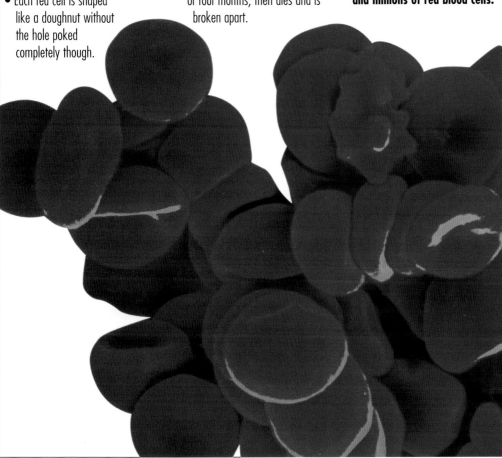

BLOOD FLOW AT REST AND AT WORK

Part of the body	Flow in millilitres per minute	
	At rest	During hard exercise
Heart	250	750
Kidneys	1,200	600
Main muscles	1,000	12,000
Skin	400	2,000
Stomach, guts, etc.	1,400	600
Brain	750	750

Only the brain's blood flow stays the same however active the body is, from running a fast race to fast asleep.

IN ONE DROP OF BLOOD

A pinhead-sized drop of blood, one cubic millimetre, contains:

- 5 million red blood cells.

- 5,000 white blood cells.

- 250,000 platelets.

If a person is ill, the number of germ-fighting white cells in blood may rise from 5,000 to 25,000.

WHAT IS IN BLOOD?

pie chart: red blood cells, plasma, other

Blood is mostly made up of plasma and red blood cells. White blood cells and platelets make up a tiny proportion of the total.

Plasma
- Forms just over half of blood by volume.
- Pale straw colour.
- Over nine-tenths is water.
- Contains many dissolved substances such as blood sugar (glucose), hormones, body salts and minerals, unwanted wastes such as urea, disease-fighting antibodies, and dozens of others.

Red blood cells
- Form just under one-half of blood.
- Also called erythrocytes.
- Carry life-giving oxygen from the lungs all around the body.
- Pick up waste carbon dioxide to take back to the lungs for removal.

White blood cells
- Form less than $\frac{1}{100}$th of blood.
- Also called leucocytes.
- Attack, disable and kill invading germs.
- Engulf or eat waste bits such as pieces of old, broken-down cells.

Platelets
- Form less than $\frac{1}{100}$th of blood.
- Also called thrombocytes.
- Are not so much whole living microscopic cells, more parts or fragments of cells.
- Help blood to clot (thrombose) to seal cuts and wounds.

WHERE IN THE BODY?

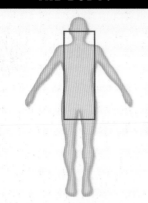

The digestive system starts at the mouth and ends at the anus. Most of its main parts, the stomach and intestines, are in the abdomen (lower half of the main body).

APPENDIX: A PUZZLING PART

The appendix is a finger-sized part of the body, branching from the start of the large intestine.

- It is a dead-end with its tip sealed, leading nowhere.

- Hollow inside.

- Varies in length from 5 to 15 cm.

- Seems to have no important task in digestion (or anything else).

The appendix may swell up with 'stuck' food and germs, causing appendicitis with severe pain in the lower right abdomen.

DIGESTION

A car needs petrol, a truck uses diesel, and a jet plane runs on kerosene. These are all fuels that provide energy to make machines go. Your 'body machine' needs fuel too – your food. It gives you the energy to move about, walk and run, and keep your inside processes going, like heartbeat and breathing. But food gives you more than energy. It provides nutrients for growth, making newer and bigger body parts, repairing old worn-out ones, and staying healthy too. The parts of the body specialised to take in and break down foods into tiny pieces are called the digestive system.

FOOD'S JOURNEY

FOOD'S JOURNEY

Part of tract	Length (cm)	Time spent by food
Mouth	10	Up to 1 minute
Throat	10	2-4 seconds
Gullet	25	2-5 seconds
Stomach	25	3-6 hours
Small intestine	580	2-4 hours
Large intestine	150	5-10 hours
Rectum	20	5-8 hours

RECYCLING

- The digestive system makes more than 10 litres of digestive juices each day.

- Most of the water in these juices is taken back into the body by the large intestine.

- Only about 0.1 litres (100 millilitres) is lost in the wastes from the system.

- The digestive system recycles 99/100ths of its water.

STOMACH

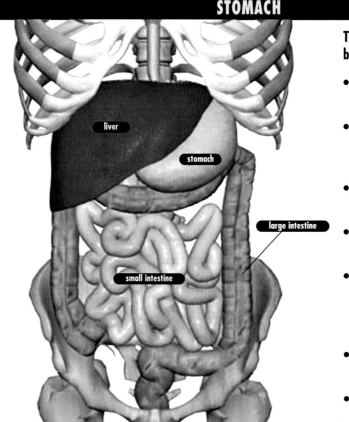

liver

stomach

large intestine

small intestine

The stomach is a J-shaped bag behind the left lower ribs.

- Measures about 30 cm around its longer side.

- Has thick muscle layers in its walls that squirm and squeeze to mash the food inside.

- Average amount of food and drink contents 1.5 litres.

- Lining makes about 1.5 litres of gastric juices each day.

- Gastric juices include hydrochloric acid, and digestive chemicals called enzymes – pepsin attacks proteins in food, and lipase attacks fats.

- Takes in or absorbs few nutrients, including sugars.

- Lining also makes thick slimy mucus, to protect the stomach's gastric juices from digesting itself.

THE DIGESTIVE TRACT

The digestive system includes the digestive passageway or tract described here, and also parts which work along with this, including the liver and pancreas. The digestive tract is the 'tube' or passageway for food.

Mouth
Where teeth bite and chew food, the tongue moves it around, and salivary glands add saliva (spit) to it.

Throat
Where swallowed food from the mouth passes down to the gullet.

Stomach
Where food is mashed and mixed with powerful juices containing acids and enzymes.

Gullet
The gullet, also known as the oesophagus or food pipe, carries food through the neck and chest to the stomach.

Small intestine
Receives part-digested, soupy food from the stomach, adds more juices and enzymes to it, and soaks up or absorbs the various nutrients and goodness.

Large intestine (colon)
Takes the leftovers from the small intestine, takes in most of the water from them, and forms them into brown squishy lumps.

• See pages 42-43 for information on the liver and pancreas.

Rectum
Stores the brown lumpy leftovers and wastes.

Anus
The last part of the passageway, a ring of muscle that relaxes to let out the brown lumpy leftovers and wastes.

• See page 29 for information on saliva.

SMALL INTESTINE

- Average width 3-4 cm.
- Has three main parts: duodenum (25 cm), jejunum (225 cm) and ileum (300 cm).
- Receives digestive juices from the pancreas and liver.
- Inner surface has many folds called plicae.
- On these folds are tiny finger-like shapes, villi, about one millimetre long.
- All the villi joined end to end would stretch 400 km.
- On each villus are even more microscopic finger-like shapes, microvilli.
- Plicae, villi and microvilli greatly increase surface area of inside of small intestine, to about 5-10 sq m, to absorb as many nutrients as possible from food.

There are about 500 million villi in the body.

PUSHING FOOD

- Without food inside, most of the digestive passageway would be squeezed flat by the natural pressure of parts or organs inside the body.
- Food has to be pushed through the passageway by waves of muscle action in its walls, called peristalsis.

In the gullet, peristalsis is so strong it works even if the body is upside down.

LEFTOVERS

- The average weight of waste (bowel motions or faeces) is 150 grams per day.
- The weight varies greatly, increasing with the amount of fibre in food.

Two-thirds water, about 100 grams (100 millilitres).

One-ninth is dead 'friendly' microbes that live naturally in the large intestine and help with digestion.

One-ninth or more of the rest is undigested food, especially fibre.

One-ninth is rubbed-off parts of the digestive lining (cheeks, throat, gullet, stomach, intestines).

• See page 41 for information on fibre.

LARGE INTESTINE

- Average width 6-7 cm.
- First part is wider, the caecum, and has the appendix branching from it.
- Second part is the ascending colon, going up the right side of the abdomen.
- Third part is the transverse colon, across the top of the abdomen.
- Fourth part is the sigmoid (S-shaped) colon, on the lower left of the abdomen.
- Mainly takes back water and valuable body salts and minerals from leftover foods and wastes.

DIFFERENT SORTS OF FOOD

Food can be divided into groups. This can help us plan a balanced diet.

Yellow: meat and fish
Green: cereals
Red: fruit and vegetables
Blue: dairy products
Orange: sugary foods

We should eat foods from each of these groups in order to stay healthy. We should choose more cereals, fruit and vegetables, and eat less sugary food.

MINERAL CHART

Iron
Needed for: Red blood cells, skin, muscles, resisting stress, fighting disease.

Calcium
Needed for: Bones, teeth, nerves, heartbeat, blood clotting, kidneys.

Sodium
Needed for: Nerves and nerve signals, digestion, blood, chemical processes inside cells, kidneys.

Magnesium
Needed for: Heartbeat and rhythm, energy use.

Iodine
Needed for: Overall speed of body's chemical processes, making thyroid hormone.

Zinc
Needed for: Healthy immune system, wound healing, maintaining senses of smell and taste.

FOOD & NUTRIENTS

Have you had plenty of lipids, complex carbohydrates, trace metallic elements and cellulose today? These are all substances that your body needs to stay healthy, but their names are quite complicated. Usually we call these substances, and everything else the body needs to take in every day, by a simpler name – food. However in your food there are six main groups of substances which you may have heard about: proteins, carbohydrates, fats, vitamins, minerals and fibre.

DAILY NEEDS

For an average adult with a typical job (not too inactive, nor very active), daily needs in grams are:

Carbohydrates   300
(including 25 grams or more of fibre)

Proteins   50

Fats   60
(mostly from plant sources)

Vitamins, examples:
Vitamin C 0.06 grams
Vitamin K 0.00008 grams

Minerals, examples:
Calcium   1
Iron   0.018
Chloride   3.4

CARBOHYDRATES

Carbohydrates are 'energy foods'.

- Broken down by digestion into smaller, simpler pieces such as sugars.

- Taken in and used as the body's main source of energy, called glucose or blood sugar, for the life processes of all its microscopic cells.

- Also used for all muscle movements from heartbeat and breathing to fast running.

- Foods with plenty of carbohydrates are starchy or sweet such as rice, wheat, barley and other cereals or grains, products made from these like pasta and bread, also many 'starchy' vegetables like potatoes, swedes, turnips and sweetcorn, and many sweet fruits like bananas and strawberries, as well as sugary items like sweets and chocolate.

Bread is made from wheat, a good source of carbohydrate.

FATS AND OILS

Fats are used for both building and energy.

- Broken down by digestion into smaller, simpler pieces called lipids.

- Foods with plenty of fats include red meats, oily fish, eggs, milk and cheese and other dairy products.

- Some plant foods also contain fats, such as avocados, olives, peanuts and soya.

- Taken in and used both for building parts such as nerves, for making and repairing parts of microscopic cells, and for energy if carbohydrates are lacking.

- Too much fat from animal sources (fatty red meats, and processed foods like burgers and salami) is linked to various health problems such as heart disease and high blood pressure.

Fat is an important part of our diet, but we should get most of what we need from plant sources, such as olive oil.

PROTEINS

Foods rich in protein are sometimes called 'building foods' as they help to build our body parts.

- Broken down by digestion into smaller, simpler pieces called amino acids.

- Amino acids are taken in and built back up into the body's own proteins which make up the main structure of muscles, bones, skin and most other parts.

- Foods with plenty of protein include all kinds of meats, poultry like chicken, fish, eggs, dairy products like milk and cheese, also some plant foods like nuts, soybeans and other beans and peas.

Dairy products such as cheese and sour cream are a good source of protein.

FIBRE

Fibre is needed to help food pass through our body properly.

- Sometimes called 'roughage'.

- Is not really broken down or digested by the body, but passes through the digestive tract largely unaltered.

- Needed to give food 'bulk' so the intestines can grip it and make it move along through them properly.

- Helps to satisfy hunger, reducing the temptation to eat too much.

- Reduces the risk of wastes being too small and hard and getting 'stuck', called constipation.

- Fibre also reduces risks of various intestinal diseases including certain kinds of colon cancer.

Fresh fruit and vegetables are a good source of fibre, particularly if the the skins are eaten as well.

FIVE-A-DAY GUIDE

'Five-a-day' means five helpings, portions or servings of fresh fruits or vegetables each day.

This should provide the body with enough of all the vitamins and minerals, as well as some energy and plenty of fibre.

Drinking a glass of fresh fruit juice counts as one portion of your 'five-a-day' target.

VITAMINS

- Needed for various body processes to work and stay healthy, and ward off disease.

- Most are needed in small amounts, fractions of a gram per day.

- Have chemical names and also letters like A, B and so on.

- The body can store some vitamins but needs regular supplies of others.

- Eating a wide range of foods, especially fresh fruits and vegetables, should provide all the body's needs.

Snacking on fresh fruit and vegetables will help to meet your body's vitamin requirements.

MINERALS

- Needed for various body processes to work and stay healthy, and ward off disease.

- Most are needed in small amounts, fractions of a gram per day.

- Most are simple chemical substances, metal elements, like iron, calcium and sodium.

- The body can store some minerals but needs regular supplies of others.

- Eating a wide range of foods, especially fresh fruits and vegetables, should provide all the body's needs.

VITAMIN CHART

Vitamin A
Chemical name: Carotene
Needed for: Eyes, skin, teeth, bones, general health.

Vitamin B1
Chemical name: Thiamine
Needed for: Brain, nerves, muscles, heart, energy use, general health.

Vitamin B2
Chemical name: Riboflavin
Needed for: Blood, eyes, skin, hair, nails, fighting disease, general health.

Vitamin B3
Chemical name: Nicotinic acid
Needed for: Energy use, controlling blood contents.

Vitamin B6
Chemical name: Pyridoxine
Needed for: Chemical processes inside cells, brain, skin, muscle, energy use, general health.

Vitamin B12
Chemical name: Cobalamin
Needed for: Blood, brain, nerves, growing, energy use, general health.

Vitamin BF
Chemical name: Folic acid
Needed for: Blood, digestion, growth.

Vitamin C
Chemical name: Ascorbic acid
Needed for: Teeth, gums, bones, blood, fighting disease, skin, general health.

Vitamin D
Chemical name: Calciferol
Needed for: Bones, teeth, nerves, heart, others general health.

Vitamin E
Chemical name: Tocopherol
Needed for: Blood, cell processes, muscles, nerves, general health.

Vitamin K
Chemical name: Phylloquinone
Needed for: Blood clotting, general health.

LIVER & PANCREAS

Your body can't digest food with just its digestive tract (passageway) – mouth, gullet, stomach and intestines. Also needed are two parts called the liver and pancreas. These are next to the stomach and they are digestive glands, which means they make powerful substances to break down food in the intestines. Together with the digestive tract, the liver and pancreas make up the whole digestive system.

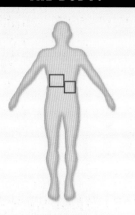

The liver is in the upper abdomen, behind the lower right ribs. The pancreas is in the upper left abdomen, behind the stomach.

WARM LIVER

The liver is so busy with chemical processes and tasks that it makes lots of heat.

• When the body is at rest and the muscles are still, the liver makes up to one-fifth of the body's total warmth.

• The heat from the liver isn't wasted. The blood spreads out the heat all around the body.

See pages 34-35 for information on the circulatory system.

GALL BLADDER AND BILE

liver

pancreas

gall bladder

THE LIVER'S TASKS

The liver has more than 500 known tasks in the body – and probably more that haven't yet been discovered. Some of the main ones are:

• Breaking down nutrients and other substances from digestion, brought direct to the liver from the small intestine.

• Storing vitamins for times when they may be lacking in food.

• Making bile, a digestive juice.

• Breaking apart old, dead, worn-out red blood cells.

• Breaking down toxins or possibly harmful substances, like alcohol and poisons.

• Helping to control the amount of water in blood and body tissues.

Alcohol is a toxin which the liver breaks down and makes harmless. Too much alcohol can overload the liver and cause a serious disease called cirrhosis.

• If levels of blood sugar (glucose) are too high, hormones from the pancreas tell the liver to change the glucose into glycogen and store it.

• If levels of blood sugar (glucose) are too low, hormones from the pancreas tell the liver to release the glycogen it has stored.

The gall bladder is a small storage bag under the liver.

• It is 8 cm long and 3 cm wide.

• Some of the bile fluid made in the liver is stored in the gall bladder.

• The gall bladder can hold up to 50 millilitres of bile.

• After a meal, bile pours from the liver along the main bile duct (tube), and from the gall bladder along the cystic duct, into the small intestine.

• Bile helps to break apart or digest the fats and oils in foods.

• The liver makes up to one litre of bile each day.

HOW THE PANCREAS WORKS

Fatty foods, such as chips, are broken apart by enzymes made in the pancreas.

- Pancreas has two main jobs.

- One is to make hormones.

- The other is to make digestive chemicals called pancreatic juices.

- These juices contain about 15 powerful enzymes that break apart many substances in foods, including proteins, carbohydrates and fats.

- Pancreas makes about 1.5 litres of digestive juices daily.

- During a meal these pass along the pancreatic duct tubes into the small intestine, to attack and digest foods there.

• See page 52 for information on hormones.

WHEN THINGS GO WRONG

A yellowish tinge to the skin and eyes is known as jaundice, and it is often a sign of liver trouble.

Usually the liver breaks down old red blood cells and gets rid of the colouring substance in bile fluid. If something goes wrong the colouring substance builds up in blood and skin and causes jaundice. Hepatitis, an infection of the liver, can cause jaundice.

UNUSUAL SUPPLY

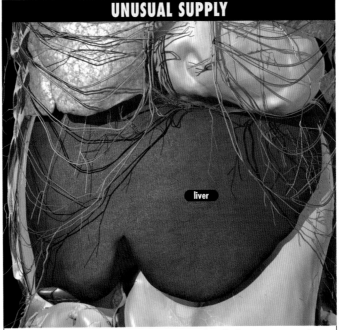

liver

One of the liver's main functions is to break down nutrients for the body. This means the liver has a unique blood supply.

- Most body parts are supplied with blood flowing along one or a few main arteries.

- The liver has a main artery, the hepatic artery.

- The liver also has a second and much greater blood supply.

- This comes along a vessel called the hepatic portal vein.

- The hepatic portal vein is the only main vein that does not take blood straight back to the heart.

- It runs from the intestines to the liver, bringing blood full of nutrients from digestion.

• See pages 36-37 for information on the blood.

BABY LIVER

Most babies and young children have big tummies (abdomens). This is partly because their liver is much larger in proportion to the body's overall size, than the liver of an adult.

- An adult liver is usually $\frac{1}{40}$th of total body weight.

- A baby's liver is nearer $\frac{1}{20}$th of total body weight.

By the time a baby becomes a toddler, their liver isn't such a large proportion of their total body weight.

WHAT IS THE LIVER?

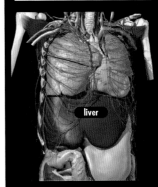

liver

The liver is the largest single part or organ inside the body.

- Wedge-shaped, dark red in colour.

- Typical weight 1.5 kg.

- Depth at widest part on right side 15 cm.

- Has a larger right lobe and smaller left lobe.

- Lobes separated by a strong layer, the falciform ligament.

WHAT IS THE PANCREAS?

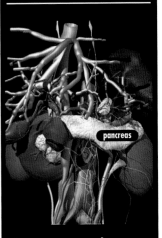

pancreas

The pancreas is a long, slim, wedge- or triangular-shaped part.

- It is soft, greyish-pink in colour.

- Typical weight 0.1 kg.

- Typical length 15 cm.

- Has three main parts: head (wide end), body (middle) and tail (tapering end).

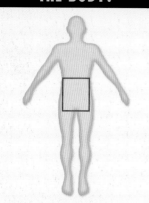

The parts of the urinary system are inside the lower body or abdomen. The kidneys are at the upper rear, on either side of the backbone behind the lower ribs. The bladder is the lowest part of the abdomen, between the hips.

MAIN PARTS OF THE KIDNEY

Cortex
Outer layer, formed mainly of microscopic blood vessels from the filtering units or nephrons.

Medulla
Inner layer, formed mainly from the tiny tubes of the filtering units or nephrons.

Renal pelvis
Space in the middle of the kidney where urine collects.

KIDNEYS & URINARY SYSTEM

You are a bit of a waste – that is, your body makes wastes and unwanted substances, which it must remove. Some of these wastes come out of the end of the digestive tube and are often called solid wastes. The other main kind is liquid wastes, sometime called 'water', but with the proper name of urine. This is very different from solid wastes. It is made, not from digestive leftovers, but by filtering the blood.

• See pages 38-39 for information on digestion.

KIDNEYS – SIZE AND SHAPE

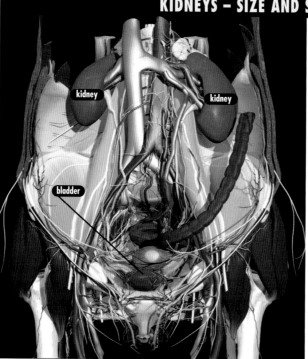

kidney

kidney

bladder

• The right kidney is usually about 1-1.5 cm lower than the left one.

• Each kidney is shaped like a bean (or kidney!) with a slight hollowing or indent on one side.

• Average kidney measurements are 11 cm high, 6 cm wide and 3 cm from front to back.

• In a woman, the typical weight of the kidneys is 130-140 grams.

• In a man, the typical weight is 140-150 grams.

The kidneys' huge blood supply can be seen by the size of the renal arteries (red) and veins (blue). The pale ureter tubes lead down to the bladder.

KIDNEY MICROFILTERS

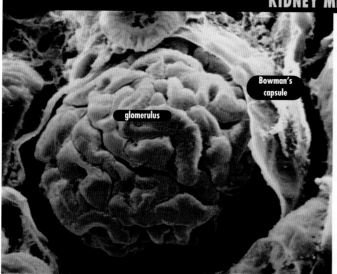

Bowman's capsule

glomerulus

Each kidney contains about one million microscopic filters called nephrons.

• Each nephron begins with a tiny tangle or knot of the smallest blood vessels, capillaries, known as the glomerulus.

• Waste substances and water squeeze out of the glomerulus into a cup-shaped part around it, Bowman's capsule.

It is in the nephrons where the work of the kidney takes place.

• The wastes and water then flow through a microscopic tube, the renal tubule, where some water, minerals and salts are taken back into the blood.

• All the tiny tubules of all the nephrons in one kidney, straightened out and joined end to end, would stretch 100 km.

• The end result is urine, which is mostly water containing dissolved wastes like urea and ammonia.

BLADDER – THE NEED TO GO

kidney

ureter

kidney

ureter

bladder

When empty, the bladder is pear-shaped and not much bigger than a thumb. It gradually stretches and fills with urine until it is emptied.

- We can tell how much urine is inside the bladder by how much we need to urinate.

- 250-300 millilitres of urine (about the amount in a coffee mug) — slight urge to urinate.

- 400-500 millilitres — stronger urge to urinate.

- 500-600 millilitres — desperate urge to urinate.

Bacteria can infect the bladder, causing cystitis. Symptoms include a burning sensation during urination and an urgent need to empty the bladder, although little urine comes out.

FEMALE AND MALE

- The urethra, which takes urine from the bladder to the outside, is different in females and males.

- This is because in males the urethra is part of the reproductive system as well as the urinary system.

- In females the urethra is 4 cm long and 6 mm across.

- In males it is 18 cm long and runs along the inside of the penis.

Men and women have different systems for taking urine to the outside of the body.

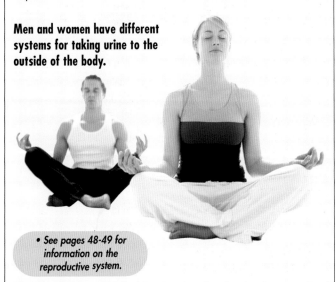

• *See pages 48-49 for information on the reproductive system.*

MAIN URINARY PARTS

Kidneys
Filter the blood to make waste liquid, known as urine.

Ureters
Tubes about 25-30 cm long which carry urine from the kidneys to the bladder.

Bladder
Properly called the urinary bladder, it stores urine until it is 'convenient' to get rid of it.

Urethra
Tube which carries urine from the bladder to the outside.

MEAT EATER

When you eat lots of meat, your urine gets darker. This is because your body makes urea, which gives urine its colour, from protein.

• *See pages 40-41 for information on the diet.*

BLOOD TO URINE

The kidneys receive more blood, for their size, than any other body part.

Amount of blood
- Each minute at rest, the kidneys receive 1.2 litres of blood.

- This is about one-fifth of all the blood pumped out by the heart.

Quick flow
- This blood flows quickly through the kidneys, so they don't actually contain one-fifth of all the body's blood.

- Over 24 hours, all the blood in the body passes through the kidneys more than 300 times.

Amount of urine
- From this blood is filtered, on an average day, about 1.5 litres of urine.

Variation of amount
- However the amount of urine varies according to how much water is taken into the body in foods and drinks.

- The amount of urine also varies according to how much water is lost in hot conditions as sweat.

Hot and cold
- On a hot day with few drinks, urine volume may be less than 1 litre.

- On a cold day with many drinks, urine volume may be more than 5 litres.

See pages 34-35 for information on the circulatory system.

GENETICS

Most children look like their parents – and this is because of genetics. Genes are instructions for how a human body grows, develops, maintains and repairs itself, and usually keeps itself healthy. Genes are passed on, or inherited, from parents to children. Instructions for building a mechanical machine are usually written as words and drawn as diagrams. But the genetic instructions for building and running the 'human machine' are in the form of a chemical substance called DNA.

Genes, in the form of the chemical DNA, are present in almost all the microscopic cells in the body. Only a few cell types, like red blood cells, lack them.

DNA

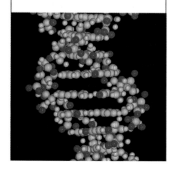

DNA is de-oxyribonucleic acid. It is a chemical substance shaped like two ladders held together and twisted like a corkscrew, which is called a double-helix.

- DNA contains four kinds of chemical sub-units called bases, named adenine(A), thymine (T), guanine (G) and cytosine (C).

- Like words in a sentence, these bases are in a certain order along DNA.

- The order of the bases is a code, the genetic code, carrying the genetic instructions.

- In the full set of DNA there are 3,200 million sets of bases.

- This full set of DNA containing all the genes for the human body is called the human genome.

CHROMOSOMES

- There are 46 lengths of DNA in each cell in the body, in the cell's control centre or nucleus.

- Each length is tightly wound or coiled into a shorter, thicker item called a chromosome.

- The chromosomes are not all different, they are in 23 pairs.

- One chromosome from each pair came from the mother, and one from the father.

- Each time a cell divides as part of growth and normal body maintenance, all the chromosomes are copied.

- So each of the two resulting cells has the complete set of 23 pairs.

- All 46 pieces of DNA in the chromosomes from a single cell, straightened out and joined together, would stretch almost two metres.

- If the same was done to all the DNA in the body, it would stretch from the Earth to the Sun and back 100 times!

GENES IN LIVING THINGS

The genes that make a mouse are almost the same as those that make a person. Out of every 100 genes, a mouse has 92 that are the same as ours.

All humans have the same overall set of genes. Tiny differences, in less than 1 out of 500 genes, make each of us unique.

- The exact number of genes that humans have is still disputed.

- Scientists think people have somewhere between 30,000 and 35,000 genes.

- Nearly all other living things have genes formed of DNA, like humans.

- Often the same genes doing the same jobs occur in very different kinds of living things.

- The more similar living things are to us, the more similar the sets of genes.

- A chimp has 98 out of every 100 genes the same as us.

- A fruit-fly has 44 out of every 100 genes the same as us.

- The tiny fungus (mould) called yeast has 26 out of every 100 genes the same as us.

STRONG AND WEAK GENES

In any family, the children usually look like at least one of their parents. Sometimes children look more like their grandparents, because some characteristics do not show up in every generation.

- Each gene can exist in several forms or versions, called alleles.

- The blue allele for the eye colour gene tells the body how to make blue colouring substance for the iris.

- The brown allele for the eye colour gene tells the body how to make brown colouring substance for the iris.

- As mentioned above, each person has two copies of a gene, one inherited from the mother, one from the father.

- A person with two alleles for brown eyes has brown eyes.

- A person with two alleles for blue eyes has blue eyes.

- If a person has one allele for blue eyes and one for brown, the brown is stronger or dominant, while the blue is weaker or recessive, and the person has brown eyes.

- Many genes work in this way, with different versions or alleles, which are dominant or recessive when put together.

- See page 49 for information on making eggs and sperm.

A gene is a portion of DNA containing the chemical code for making a part of the body, or instructing how that part works.

For example, the gene for eye colour tells the body how to make the coloured substance, or pigment, for the coloured part of the eye called the iris.

Number
- The human body has a total of about 30,000 genes.

- Sometimes several genes work together to control one feature.

Instructions for appearance
Genes instruct for skin colour, hair colour and type, overall adult height, ear lobe shape and many other features of the body.

Instructions for processes
Genes also control how the body's chemical processes work inside, like digesting food.

WHAT ARE CLONES?

Usually, each person has a unique selection of genes, possessed by no one else.

The exception is identical twins, who have exactly the same genes.

Living things with exactly the same genes are called clones.

Dolly the sheep is the most famous clone. Scientists took genes from an adult sheep and used them to create an identical copy – Dolly.

GENETIC FINGERPRINTING

The DNA from skin, hair and blood can help the police in their enquiries. Genetic information can eliminate a suspect, or help the police to secure a conviction.

Police procedures have been revolutionised since reliable DNA fingerprinting has been available.

- Small pieces of DNA, for example, from the white blood cells in a tiny speck of blood, can be copied millions of times.

- This is done by a laboratory process called PCR, polymerase chain reaction.

- PCR gives enough DNA for testing, to look at various sets or sequences of genes.

- The main testing method is called gel electrophoresis.

- The results are flat layers of a jelly-like substance containing dark stripes or bands, like a supermarket bar code.

- The sequence of bands gives a genetic fingerprint.

- If two samples of DNA match exactly, the chances are millions to one that they came from the same body.

REPRODUCTION

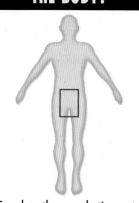

Female – the reproductive parts or sex organs are near the base of the lower body (abdomen).

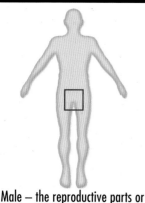

Male – the reproductive parts or sex organs are mostly below the abdomen, between the legs.

One of the main features of life is that all living things make more of their kind, by breeding or reproduction. This happens in humans too. The process of reproduction happens in the same basic way in the human body as it does in other animals like cats, dogs, horses, mice and whales. A female and male get together and have sex, which makes a tiny egg start to grow inside the female. The body parts which do this are called the sexual or reproductive system.

FEMALE PARTS

Ovaries
Every 28 days or so one egg cell is released as a ripe egg or ovum.

Oviducts
Also called fallopian tubes or egg tubes. These carry the ripe egg towards the womb.

Womb (uterus)
Where the new baby grows and develops from a fertilised egg, during pregnancy.

Cervix
This is the neck of the womb. It stays closed during pregnancy, then opens at birth to allow the baby to be born.

Vagina (birth canal)
The new baby passes along this from the womb to the outside world at birth.

EGG CELLS

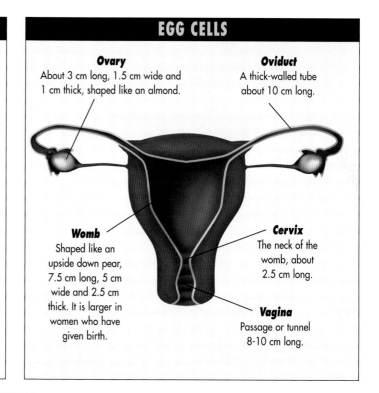

Ovary
About 3 cm long, 1.5 cm wide and 1 cm thick, shaped like an almond.

Oviduct
A thick-walled tube about 10 cm long.

Womb
Shaped like an upside down pear, 7.5 cm long, 5 cm wide and 2.5 cm thick. It is larger in women who have given birth.

Cervix
The neck of the womb, about 2.5 cm long.

Vagina
Passage or tunnel 8-10 cm long.

EGG RELEASE CYCLE

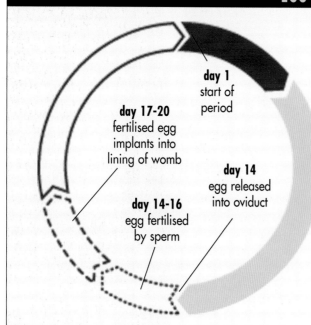

day 1
start of period

day 17-20
fertilised egg implants into lining of womb

day 14-16
egg fertilised by sperm

day 14
egg released into oviduct

The menstrual cycle lasts about 28 days. The cycle first begins when a girl is about 10 to 14 years old.

• Once every 28 days or so, an egg cell ripens and is released from its ovary, into the oviduct.

• This process is called ovulation.

• As it happens the womb lining has become thick and rich with blood, ready to nourish the egg cell if it joins with a sperm cell and begins to develop into a baby.

• If this does not happen the womb lining breaks down and is lost through the vagina as menstrual bleeding (a period).

• Then the whole process of egg ripening and womb changes starts again.

• The process is called the menstrual cycle and lasts about 28 days.

• It is controlled mainly by hormones called sex hormones, oestrogen and progesterone.

• See pages 50-51 for information on the stages of life.

• See page 53 for information on hormones.

SPERM CELLS

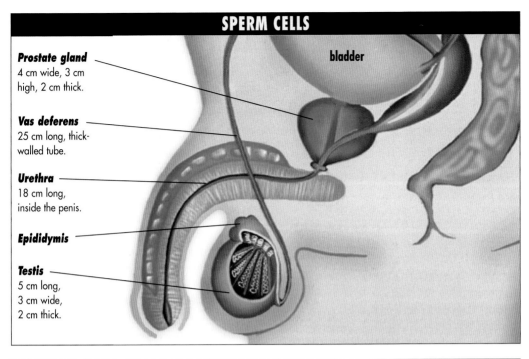

Prostate gland
4 cm wide, 3 cm
high, 2 cm thick.

Vas deferens
25 cm long, thick-
walled tube.

Urethra
18 cm long,
inside the penis.

Epididymis

Testis
5 cm long,
3 cm wide,
2 cm thick.

bladder

MALE PARTS

Penis
Contains the urethra tube along
which sperm pass as they leave the
body, at the time called ejaculation.

Testes
Make millions of microscopic sperm
cells every day.

Vas deferens
Also known as the ductus deferens or
sperm tubes. Carry sperm cells from
the testes and epididymis.

Epididymis
Store sperm cells until they
are released.

Scrotum
Bag of skin containing the testes
and epididymis.

Prostate gland
Makes nourishing fluid for the
sperm cells.

MAKING EGGS AND SPERM

**When the body makes eggs
or sperm, cells are copied in
a unique way.**

- Ordinary body cells divide into
  two to form more cells for growth,
  body maintenance and repair.

- During this process all the genes in
  all the 23 pairs of chromosomes
  are copied, so each resulting cell
  receives a full set of 23 pairs.

- This type of cell division is called
  mitosis.

- Egg and sperm cells are made
  by a different type of cell division
  called meiosis.

- In meiosis, the genes are not copied.

- Each egg or sperm receives just one
  of each pair of chromosomes,
  making 23 instead of 46.

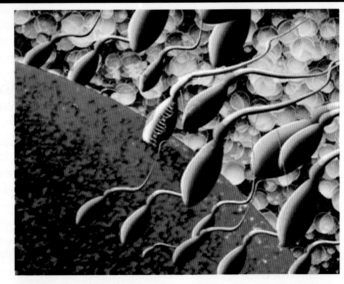

- When an egg joins with a sperm
  to start a new baby, the two
  sets of 23 chromosomes come
  together to form 23 pairs, which
  is back to the normal number.

*• See pages 46-47 for
information on genetics.*

**Genes are carried by sperm
and eggs. Each carries half of
the genetic material needed
to form a new baby. When
they meet, the halves join
together to make a new
individual.**

• See pages 46-47 for
information on genetics.

F E M A L E E G G S

**Egg cells are about 0.1 mm
across, almost microscopic.**

- At birth a new baby girl has half
  a million unripe egg cells in her
  ovaries. The number decreases
  as she gets older.

- By the time a girl has grown up
  and is ready to have children, the
  number of egg cells in her body
  is about 200,000.

- Over the years when she can
  have children, a woman's ovaries
  release about 400 egg cells.

M A L E S P E R M

**Sperm are shaped like tiny
tadpoles, with a rounded
head and long whippy tail.**

- Among the smallest cells in the
  body, just 0.05 mm in total length.

- Tens of millions are made every
  day in a massive tangle of tubes
  in the testis, called seminiferous
  tubules.

- All the tubules from one testis
  straightened out and joined end
  to end would stretch over 100 m.

- Each sperm cell takes about two
  months to form.

- Sperm are then stored in the
  epididymis tube, which is folded
  and coiled next to the testis.

- Opened out straight, the
  epididymis tube would stretch 6 m.

- When sperm are released, about
  200-500 million pass in fluid
  along the vas deferens and
  urethra, and out of the end of
  the penis.

- If they are not released, they
  break down and their parts
  are recycled within the body.

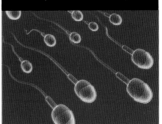

PREGNANCY

The development of a baby inside the mother's womb is called pregnancy and lasts about 9 months (average 266 days from fertilisation to birth).

GROWTH RATES

Growth is fastest during early weeks in the womb and slows down towards birth.

- Speeds up slightly in first 2-3 years after birth.
- Slows down towards end of childhood.
- Sudden spurt during puberty, usually early teens.
- Slows down towards late teens.
- Full height usually by 20 years of age.

STAGES OF LIFE

In the beginning, every human body was a tiny speck smaller than the dot on this 'i' – a fertilised egg cell. By the process of cell division (splitting), that single cell became two, four, eight and so on. About 20 years later, by the time of adulthood, the body has 50 million million (50 trillion) cells of more than 200 different kinds. It is a fascinating story of amazing growth and development. At which stage are you?

IN THE WOMB: WEEK ONE

Egg cell is joined or fertilised by sperm cell, usually in the oviduct (fallopian tube) of the mother. The genes of egg and sperm (in the form of DNA) come together and the genetic blueprint for a new body is formed.

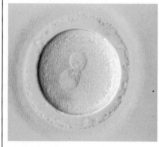

24-36 hours
Fertilised egg cell splits into two smaller cells.

36-48 hours
Each of the two cells divides, forming four cells.

2-3 days
Cell division continues, forming a tiny ball of more than a 20 cells moving slowly along the oviduct.

4 days
The ball of more than 100 cells, called a morula, reaches the inside of the womb.

5 days
The ball of hundreds of cells becomes hollow inside, called a blastocyst (early embryo).

6-7 days

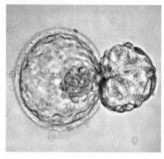

The blastocyst, still only 0.1-0.2 mm across, settles or implants into the blood-rich lining of the womb.

The cells now take in nourishment from the lining and enlarge between divisions, so the early embryo starts to grow.

IN THE WOMB: EMBRYO

For the first 8 weeks after fertilisation, the developing baby is called an embryo.

Week 2
The embryo becomes the shape of a flat disc, surrounded by fluid, within the womb lining. The disc lengthens and curls over at the edges.

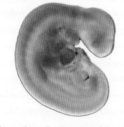

Week 3
The curled-over disc becomes longer and larger at one end, which begins the shape of the head and brain. Length 1.5 mm.

Week 4

The embryo becomes C-shaped. Simple tubes start to make the heart and begin pulsating. Arms and legs begin as small bulges on the body. Length 5 mm.

Week 5
Head and brain grow rapidly. Inner organs form like the stomach and kidneys. Nose begins to take shape. 'Tail' is still present. Length 8 mm.

Week 6

Heart and lungs almost formed, body becomes straighter. Eyes and ears obvious. Length 12 mm.

Week 7
Fingers and toes start to take shape. Neck becomes more visible. Tail shrinks. Length 15 mm.

Week 8

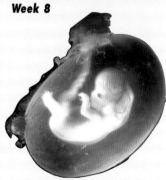

Muscles and eyelids form, tail has almost disappeared. All main body parts are present, even eyelids. Length 17 mm, about the size of a grape.

This picture shows the embryo floating in its amniotic bag of fluid.

IN THE WOMB: FETUS

For period between 8 weeks (2 months) after fertilisation and birth, the developing baby is called a fetus.

Month 3
Finishing touches are made including folds for fingernails and toenails. Eyelids joined. Head still very large compared to body. Length 40 mm.

Month 4
Face looks much more human, first hair grows on head. First bones begin to harden. Length 55 mm.

Month 5
Reproductive parts begin to take shape, showing if the fetus is a girl or boy. Length 150 mm.

Month 6
Body becomes slimmer and is covered with fine hair, lanugo. Fetus can suck thumb, move arms and kick legs,

mother feels these movements.

Month 7
Body is lean and wrinkled. Fetus can swallow, eyes can detect light.

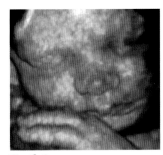

Month 8
Body puts on fat, becoming chubby. Nails grow to ends of fingers and toes.

Month 9
Baby is chubbier and fully formed. Average weight at birth 3.0-3.5 kg, average length 50-60 cm.

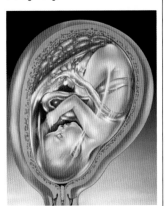

NEW BABY

By their first birthday, most babies have almost tripled their birth weight, from just over 3 to almost 10 kg. They have also grown in height from about 55 to 75 cm.

Average times for movement and coordination skills (although there are wide variations)

4-8 weeks
Smiles in response to faces

2-4 months
Raises head and shoulders when lying on tummy

5-7 months
Rolls over from tummy onto back

6-8 months
Sits up perhaps with help, starts to make babbling noises

7-9 months
Begins to try and feed itself, puts items in mouth

8-10 months
Crawls

10-12 months
Stands up with support

12-15 months
Walks unaided

SIGNS OF AGEING

Most people are at peak physical fitness from early 20s to mid 30s.

- Most men run their lifetime best between 27 and 29 years.

- Most women run their lifetime best between 29 and 31 years.

- Most marathon runners perform their best when aged between 30 and 37 years.

Certain reactions and body processes begin to slow down from the 40s in some people, but not until the 60s in others.

Signs of ageing:
- *Wrinkled skin*
- *Greying or whitening of hair*
- *Hair loss*
- *Less muscle power*
- *More brittle bones*
- *Less flexible joints*
- *Senses become less keen*
- *The first sense to deteriorate is usually hearing, followed by sight. Touch and taste also become less keen, and then smell.*

PUBERTY

Puberty is the time when the body grows and develops rapidly from girl to woman or boy to man, and the reproductive (sex) parts start to work.

Puberty - girls
- Can occur any time between 9 and 16 years of age. It usually takes 2-3 years.

- First signs include rapid growth in height, breasts begin to enlarge.

- Hair grows under arms, between legs (pubic hair).

- Hips increase in width.

- Pads of fat laid down under skin give more rounded body outline.

- Voice deepens slightly.

- Menstrual cycle begins with first period, menarche.

Puberty - boys
- Can occur any time between 11 and 17 years of age. It usually takes 3-4 years.

- First signs include rapid growth in height, hair growing under arms, between legs (pubic hair) and on face (moustache and beard area).

- Shoulders increase in width.

- Muscle development increases giving more angular body outline.

- Voice deepens considerably or 'breaks'.

- Reproductive (sex) parts enlarge and begin to make sperm.

HORMONES

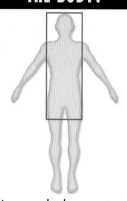

Hormone glands are scattered through the central body, from the head down through the neck into the lower abdomen. The reproductive parts known as the ovaries in women, and the testes in men, are also hormonal glands.

Your brain is boss of your body. It tells muscles to pull, the lung to breathe and the heart to beat. It does this by sending out tiny electrical signals called nerve messages. But the brain also controls certain processes in another way, by natural body chemicals called hormones. These are made in hormone or endocrine glands. The hormones spread around the body in the blood and affect how various parts work. When you are frightened and your heart pounds, and you feel 'butterflies' inside – that's a hormone called adrenaline at work.

THYROID

thyroid

PARATHYROIDS

Where
4 glands, two embedded in each side of the thyroid.

Shape
Like tiny eggs.

Size
Each is 6 mm high, 4 mm wide, 2 mm thick, weight 0.05 grams.

Hormones made
Parathyroid hormone (PTH)

Effects
Controls the level of the mineral calcium in the blood.

PITUITARY

The pituitary gland helps children to grow properly. It is sometimes called the 'master gland' because it controls several other glands.

Where
Under the skin of the neck just below the voice-box, wrapped around the upper windpipe.

Shape
Like a bow tie or butterfly.

Size
8-10 cm wide, 3 cm high, 2 cm thick, weight 25 grams.

Hormones made
Thyroxine (T4), tri-iodothyronine (T3), calcitonin

Effects
T4 and T3 make all the body's cells work faster so the whole body chemistry, or metabolism, speeds up. Calcitonin lowers the level of the mineral calcium in the blood.

Calcium is vital for our brains, muscles and blood to work properly, as well as building bones.

• See page 40 for information on calcium.

One of the smallest hormone glands, but the most important. It is controlled by the brain just above and sends out hormones which affect the workings of other hormonal glands.

Where
Behind the eyes and below the central front of the brain, joined to it by a narrow stalk.

Shape
Bean-like.

Size
1 cm high, 1.2 cm wide, 0.8 cm thick, weight 0.5 grams.

Hormones made
About 10 including growth hormone, ADH (antidiuretic hormone), and TSH (thyroid-stimulating hormone)

Effects
Growth hormone makes the whole body increase in size and development.

• ADH makes the kidneys take back more water as they form urine.

• TSH makes the thyroid gland release more of its own hormones.

• See pages 44-45 for information on the kidneys.

PANCREAS

Where
In the upper left abdomen, behind the stomach.

Shape
Long, slim, wedge- or triangular-shaped.

Size
15 cm long, about 0.1 kg in weight.

Hormones made
Insulin and glucagon, made in about one million tiny 'blobs' or cells called islets scattered through the pancreas.

Effects
Insulin lowers level of blood sugar (glucose) by making body cells take in more of it, glucagon raises the level.

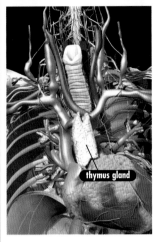

People with diabetes have a problem with the hormone insulin. They may have to inject insulin into their bodies.

> • See page 43 for information on the role the pancreas plays in digestion.

THYMUS

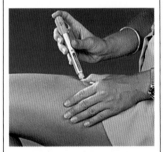

thymus gland

The thymus gland in the chest helps to make the white blood cells which destroy germs.

Where
Behind the breastbone.

Shape
Two joined sausage-like blobs or lobes.

Size
Relatively larger in young children, thumb-sized and weighing up to 20 grams, shrinks slightly during adulthood.

Hormones made
Thymosin, thymopoietin and others

Effects
Help white blood cells to develop their germ-attacking powers.

> • See pages 36-37 for information on the blood.

ADRENALS

Where
2 glands, one on top of each kidney.

Shape
Like a small pyramid or 'bonnet' on the kidney.

Size
Each is 5 cm high, 3 cm wide, 1 cm thick, weight 5 grams.

Hormones made
Outer part (called the cortex) makes corticosteroid or 'steroid' hormones including cortisol and aldosterone; inner part (called the medulla) makes adrenaline and similar hormones.

Effects
Cortisol decreases the effects of stress and helps control blood sugar and body repair, aldosterone affects how the kidneys filter blood. See box right for information on adrenaline.

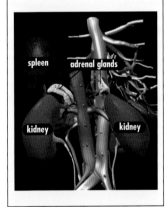

spleen   adrenal glands

kidney   kidney

FIGHT OR FLIGHT?

When the body is frightened or stressed and has to act fast, the adrenal glands releases their hormone adrenaline (also called epinephrine).

This helps it to get ready to face a threat and 'fight', or escape and flee from the danger in 'flight'. The main effects of adrenaline are:

Blood vessels
• Widens blood vessels to muscles and heart muscle.

Heart
• Increases beating rate of heart.
• Increases amount of blood pumped by each heartbeat.

Breathing
• Increases breathing rate.
• Widens airways in lungs allowing more air with each breath.

Digestion
• Decreases activity of inner parts like stomach and guts.

Sugar levels
• Raises level of blood sugar (glucose) to provide more energy for muscles.

Our ancestors would have had to act fast if they saw a potentially dangerous predator. Adrenaline prepared the body for the responses they would have made.

OTHER HORMONE-MAKING PARTS

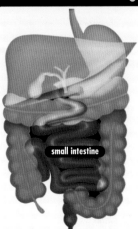

small intestine

Some body parts make hormones in addition to their main tasks.

Stomach
Gastrin makes stomach lining release acid.

Heart
Atriopeptin affects amounts of body salts and minerals, and blood pressure.

Testes
Testosterone gives men their male characteristics.

Ovaries
Oestrogen and progesterone give women their female characteristics.

> • See pages 48-49 for information on reproduction.

The small intestine makes the hormone secretin, which tells the pancreas release acid-neutralising juices.

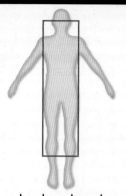

Lymph nodes and vessels are found in most body parts, especially in the neck, armpits, chest, central abdomen and groin. They are also in the adenoids at the rear of the nasal chamber, the tonsils in the throat, the spleen behind the left kidney, and the thymus gland behind the breastbone.

LYMPH & IMMUNE SYSTEMS

When people are ill, often their glands swell up – especially in the neck, armpits and groin. But these are not really 'glands'. They are called lymph nodes and they are part of the lymphatic system. This system is the body's 'alternative' circulation. Like blood, lymph fluid flows around the body in tubes, known as lymph vessels. Also like blood, it carries nutrients to many parts and collects wastes. It is closely linked to the immune system, which is specialised to attack germs and fight disease.

LYMPH NODES

There are about 500 lymph nodes all over the body.

- The larger ones are in the neck, armpits, chest, central abdomen and groin. There are others in the crook of the elbow and the back of the knee.
- Mostly shaped like balls, beans or pears.
- Smallest ones less than 1 mm across, larger ones 15-20 mm.
- Contain mainly various kinds of white blood cells.
- Can double in size when fighting illness.
- Lymph flows into each node along several lymph vessels.
- Lymph flows away from each node along one vessel.

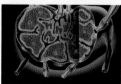

A typical lymph node contains lymph fluid and white blood cells.

LYMPH FLUID

Lymph fluid is usually pale or milky in colour.

- The average amount in the human body is 1-2 litres.
- Contains mainly water, dissolved nutrients, and disease-fighting white blood cells and the antibodies they make.
- Forms from fluid which oozes out of and between cells, and collects inside and between tissues.
- Flows slowly along smaller lymph vessels, which join to form larger ones.
- The fluid has no pump of its own (like blood has the heart) but moves by general body pressure and the massaging effect of muscles and movements.
- Flows through lymph nodes on its journey.
- Lymph network gradually gathers lymph into two main lymph vessels in the chest.
- These are the right lymphatic duct and the thoracic duct.
- These ducts join the right and left subclavian veins, where lymph joins the blood.

SPECIALIST LYMPH PARTS

ADENOIDS

- Also called pharyngeal tonsils.
- Found at the rear of the nasal chamber in the uppermost throat.
- Consist of a gathering of small lymph nodes called nodules.
- Help to kill germs in breathed-in air.
- Swell up when battling illness, causing problems with air flow through the nose, and may need removal if this happens too often.

TONSILS

Red, sore, swollen tonsils are a sign of tonsillitis.

- Also called palatine tonsils.
- Found at the sides of the throat, just below and on either side of the soft palate (rear of the roof of the mouth).
- Consist of a gathering of small lymph nodes called nodules.

- Help to kill germs in breathed-in air and also foods and drinks.
- Swell up when battling illness, causing a sore throat.
- Tonsils may need removal if this happens too often.

SPLEEN

spleen

The dark-red spleen stores blood, recycles old red blood cells and makes new white cells.

- Behind and above the left kidney.
- Largest collection of lymph tissue in the body.
- Length about 12 cm (the size of a clenched fist).
- Weight about 150 grams, but can be half or twice this according to blood content and body's state of health.

- *See page 30-31 for information on the airways.*

IMMUNE SYSTEM

When we catch a cold our immune system begins fighting it immediately. Colds usually last only a few days.

The immune system involves the lymph and blood systems and also many other body parts.

- Based on white blood cells of various kinds.

- Defends against harmful substances like toxins or poisons.

- Also protects the body against invasion by germs such as bacteria, viruses and microscopic parasites (called protists or protozoans).

- Helps to clean away bits, pieces and debris from normal body maintenance, as old cells die and break down.

IMMUNITY

Once we have caught an illness and fought the infection, we have immunity to the germ in the future.

- After the body catches an infection, especially by a virus germ, white cells called memory cells 'remember' the type of virus.

- If the germ invades again later, the immune system can recognise and fight against it straight away and usually defeat it quickly.

- This is called being resistant or immune to that particular germ.

Viruses and bacteria are contagious which means they spread between the people in close contact with each other.

TYPES OF IMMUNITY

The body becomes immune to illnesses in several ways.

Innate or native immunity
Already in the body.

Acquired immunity
Occurs after exposure to antigens, for example, on the surface of a type of germ.

Natural acquired immunity
Happens when the body catches the germ naturally.

Artificial acquired immunity
When an altered form of the germ or its products is put into the body specially, by vaccination.

Active immunity
When the body makes its own antigens.

Passive immunity
When ready-made antibodies are put into the body.

We are exposed to many germs in our every day life. As we grow up, our bodies develop immunity to most of them naturally. There are only a few diseases we need to be protected from artificially, through immunisations.

MAIN DEFENDERS – LYMPHOCYTES

Lymphocytes are one of the main kinds of white blood cells.

The healthy body contains about 2 trillion lymphocytes. They are made in bone marrow, and there are two main kinds, B-cells and T-cells.

T-CELL LYMPHOCYTES

- T-cell lymphocytes are processed or 'trained' in the thymus gland in the chest.

- T-cell lymphocytes attack and kill invading 'foreign' cells like bacteria directly.

- T-cell lymphocytes also encourage white cells called macrophages to engulf or 'eat' invading microbes.

- T-cell lymphocytes encourage B-cells to make antibodies.

B-CELL LYMPHOCYTES

- B-cell lymphocytes respond to the chemical messages from T-cell lymphocytes.

- B-cell lymphocytes are encouraged into activity by T-cells.

- B-cells change into plasma cells which make defensive substances called antibodies.

- Different antibodies are made in response to different antigens, which are 'foreign' substances on various invading microbes or made by them.

- Antibodies spread around in the blood and lymph.

- Antibodies join to antigens and make them ineffective or destroy them.

• See page 16 for information on bone marrow and page 36-37 for information on blood.

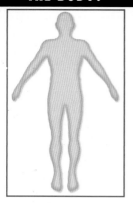

Any part of the body has the potential to stop functioning as well as it should.

Most people are well, most of the time. Nearly everyone gets the occasional cold and cough. Some of us have a few bigger health problems, like an infection such as chickenpox, or perhaps an injury such as a sprained joint or broken bone. A few people are less lucky and seem to be ill quite often. But the basic rules for good health are the same: do not smoke, eat a nutritious and balanced diet, take plenty of exercise and activity, and keep a positive approach or attitude to life.

TYPES OF MEDICINES

Anaesthetic
Reduces or gets rid of sensations including pain.

Analgesic
Reduces or 'kills' pain.

Beta-blocker
Slows heartbeat and makes it more regular, lowers blood pressure.

Bronchodilator
Widens (dilates) the small airways (bronchioles) in the lungs.

Cytotoxic
Destroys cells, especially malignant or cancerous ones, in chemotherapy.

Diuretic
Reduces water content of the body by increasing amount of urine.

Immunosuppressive
Reducing or damping down the actions of the body's immune defence system, for example, so it does not reject a body part transplanted from another person.

Steriods
Large group of medicines including hormones and substances which help to increase muscle size or damp down the action of the immune defence system.

Thrombolytic
'Clot-buster' to dissolve blood clots.

Vasoconstrictor
Making the small arterial blood vessels narrower.

Vasodilator
Making the small arterial blood vessels wider.

MEDICAL DRUGS

Many drugs have a name that begins 'anti-'. This shows which problem the drug fights against.

Antibiotic
Disables or destroys microbes, mainly bacteria.

Anticoagluant
Prevents blood clotting.

Anticonvulsant
Lessens the risk of convulsions, fits or seizures.

Antidepressant
Reduces the effects of depressive illness by lifting mood and outlook.

Anti-emetic
Reduces feelings of sickness or nausea.

Antifungal
Disables or destroys fungal microbes.

Antihistamine
Used against allergy related illnesses such as hay fever, asthma, food allergy.

Anti-inflammatory
Reducing inflammation (redness, swelling, soreness, pain).

Antipyretic
Lowers fever, reduces body temperature.

Antiseptic
Kills most kinds of germs, usually applied to outside of body (skin).

Antitoxin
Makes a poisonous or toxic substance harmless.

Antiviral
Disables or destroys virus microbes.

MAJOR CAUSES OF ILLNESS AND DISEASE

It can be quite hard for young children to describe their symptoms. This can make diagnosis difficult for the doctor.

- Genetic problems, which can be passed on from parents, or begin as a new case when genes are not copied correctly as cells multiply in the developing body.

- Congenital problems, which are present at birth.

- Infections caused by invading microbes.

- Infestations due to worms, fleas and similar animals.

- Malnutrition, due to not enough food and/or foods which are not balanced in their nutrients.

- Toxins or poisons from the surroundings, such as in water, or breathed in air.

- Physical harm by accidents and injuries.

- Radiation such as radioactivity, nuclear rays, the Sun's harmful ultra-violet rays.

- Cancers, where cells multiply out of control and spread to invade other body parts.

- Autoimmune problems, where the body's immune defence system mistakenly attacks its own parts (as in rheumatoid arthritis).

- Allergic disorders, where the body's immune defence system mistakenly attacks harmless substances, like plant pollen in hay fever.

- Metabolic conditions, when there is a problem with the inner processes of body chemistry.

BACTERIA

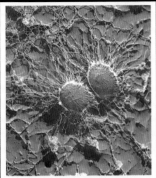

Two bacteria, magnified 50,000 times.

- Each bacteria is a single unit of life, a cell.

- Microscopic, about 100,000 would fit inside this 'o'.

- Different groups are known by their shapes such as cocci (balls or spheres), bacilli (rods or sausages) and spirochaetes (corkscrews).

- Many are disabled or killed by antibiotic drugs.

- Bacterial infections include many kinds of sore throat, skin boils, pertussis (whooping cough), tetanus, scarlet fever, most kinds of cholera and dysentery, plague, diphtheria, Legionnaire's disease, many kinds of bronchitis, most types of food poisoning like salmonella and listeria and botulism, ear infections and anthrax.

MICRO-FUNGI

- Each micro-fungus is a single unit of life, a cell.

- Belong to the fungus group which includes mushrooms, toadstools and yeasts.

- Often grow and spread on the skin.

- Diseases caused include athlete's foot, ringworm, thrush (candida), fungal nail and hair infections.

PROTISTS (PROTOZOA)

The disease malaria is caused by a protist. It is transmitted to humans through mosquitos.

- Each protist is a single unit of life, a cell.

- Sometimes called parasites or microparasites.

- Most are microscopic, about 1,000 would fit inside this 'o'.

- More common in warmer countries.

- Diseases caused include malaria (plasmodium), sleeping sickness (trypanosome), Chagas' disease (trypanosoma cruzi), river blindness (schistosomiasis).

VIRUSES

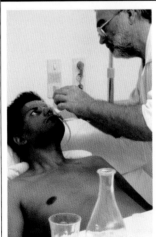

Scientists can only treat the symptoms of most viral infections, not the virus itself.

- Smallest known forms of life, if indeed they are truly alive.

- About half a million would fit on this full stop.

- Invade the body's own microscopic cells.

- Take over these cells and make them produce more copies of the virus.

- Viral infections include common colds, most types of 'flu (influenza), polio (poliomyelitis), skin warts, chickenpox, measles, mumps, rubella, yellow fever, most kinds of hepatitis, rabies, Ebola, AIDS (due to Human Immunodeficiency Virus, HIV).

- Are not affected by antibiotic drugs.

- Several viral infections can be prevented by vaccination to make the body resistant or immune to them (immunisation).

• See page 55 for information on types of immunity.

DISEASE-CAUSING MICROBES

- Several kinds of harmful microbes cause the illnesses known as infectious diseases or infections.

- They are known in everyday terms as 'germs' or sometimes 'bugs'.

- Most get into the body in breathed-in air, in foods or drinks, or through cuts and open wounds.

- When a harmful microbe spreads by direct or personal contact between people, rather than by indirect means such as air or water, this is known as a contagious infection.

- The time when the harmful microbe multiplies inside the body, but the person does not show signs of illness, is the incubation period.

MEDICAL SPECIALISTS

Anaesthetist
Giving anaesthetics to remove sensation and pain while patient is still conscious (local anaesthetic) or to make the patient unconscious as well (general anaesthetic).

Cardiologist
Heart and main blood vessels.

Dermatologist
Skin, hair and nails.

Gynaecologist
Female parts, usually sexual and urinary organs.

Geriatrics
Older people.

Haematologist
Blood and body fluids.

Neurologist
Brain and nerves.

Obstetrician
Pregnancy and birth.

Oncologist
Tumours, especially cancers and similar conditions.

Ophthalmologist
Eyes.

Orthopaedic surgeon
Bones, joints and the skeleton.

Paediatrician
Care of children.

Pathologist
The processes and changes of disease, such as laboratory studies of samples.

Physiotherapist
Using physical measures such as massage, manipulation, exercise, heat.

Psychiatrist
Mental and behavioural problems.

Radiologist
X-rays and other imaging methods.

Thoracic surgeon
Chest, especially lungs and airways, also heart.

Urologist
Urinary system of kidneys, bladder and their tubes.

GLOSSARY

Abdomen The lower part of the main body or torso, below the chest, which contains mainly digestive and excretory (waste-disposal) parts, and in females, reproductive parts.

Artery A blood vessel (tube) which conveys blood away from the heart.

Axon The very long, thin part of a nerve cell or neuron, also called a nerve fibre.

Bladder Bag-like sac or container for storing fluids. The body has several, including the urinary bladder (often just called 'the bladder') and gall bladder.

Blood sugar Also called glucose, the body's main energy source, used by all its microscopic cells to carry out their life processes and functions.

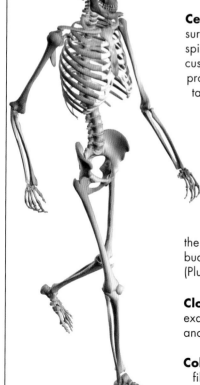

Capillary The smallest type of blood vessels, usually less than one millimetre long and too thin to see except through a microscope.

Cardiac To do with the heart.

Cartilage Tough, light, slightly bendy and compressible body substance, often called 'gristle', which forms parts of the skeleton such as the ears and nose, and also covers the ends of bones in joints.

Cell The basic microscopic 'building block' of the body, a single living unit, with most cells being 0.01-0.05 mm across. The body contains over 50 million million cells.

Central nervous system The brain and spinal cord.

Cerebral To do with the cerebrum, the largest part of the brain which forms its wrinkled domed shape.

Cerebrospinal fluid Liquid surrounding the brain and spinal cord, to protect and cushion them as well as help provide nourishment and take away wastes.

Cilium Microscopic hair, usually sticking out from the surface of a cell, which can wave or bend, and perhaps sense substances as in the olfactory epithelium of the nose and in the taste buds on the tongue. (Plural: cilia.)

Clone A living thing with exactly the same genes as another living thing.

Collagen Tiny, tough, strong fibres found in body parts such as skin and bones.

Cortex The outer part or layer of a body part, such as the renal cortex of the kidney, or the cerebral cortex of the brain.

Cranium The upper domed part of the skull or 'brain case', which covers and protects the brain.

Cermis The inner layer of skin, under the epidermis (*see below*), which contains the touch sensors, hair follicles and sweat glands.

DNA De-oxyribonucleic acid, the chemical substance that forms the genetic information or genes.

Embryo The name for a developing human body, from fertilisation as a single cell, to eight weeks later.

Endocrine To do with hormones and the hormonal system (*see hormone*).

Enzymes Substances which alter the speed of a chemical change or reaction, usually speeding it up, but which remain unchanged themselves at the end of the reaction.

Epidermis The protective outer layer of skin, which is always being worn away but continually replacing itself.

Excretory To do with removing waste substances from the body. The main excretory system is made up of the kidneys, bladder and their linking tubes.

Fertilisation When an egg cell joins a sperm cell to start the development of a new human body.

Fetus A developing human body from eight weeks after fertilisation until birth.

Fovea The small area in the retina of the eye where vision is most detailed and clearest, due to the great number of cone cells.

Gastric To do with the stomach.

Gland A body part that makes a substance or product which it then releases, such as the tear glands which make tear fluid for the eyes, and the sweat glands in the skin.

Glucose *See* blood sugar.

Gustatory To do with the tongue and taste.

Hepatic To do with the liver.

Hormone A natural 'chemical messengers' that circulates in the blood and affects how certain body parts work, helping the nervous system to control and coordinate all body processes.

Humour Old word used to describe various body fluids, still used in some cases, for example, to describe the fluids inside the eye, the vitreous ('glassy') humour and aqueous ('watery') humour.

Immunity Protection or resistance to microbial germs and other harmful substances.

Integumentary Concerning the skin and other coverings, including nails and hair.

Ligament A stretchy, strap-like part that joins the bones around a joint, so the bones do not move too far apart.

Medulla The inner or central region of a body part, such as the renal medulla of the kidney, or the adrenal medulla of the adrenal gland.

Meninges Three thin layers covering the brain and spinal cord, and also making and containing cerebrospinal fluid. They are known as the dura mater, arachnoid and pia mater.

Meiosis Part of special type of cell division, when the chromosomes are not copied and only one set (not a double-set) moves into each resulting cell.

Metabolism All of the body's thousands of chemical processes, changes and reactions, such as breaking apart blood sugar to release energy, and building up amino acids into proteins.

Mineral A simple chemical substance, usually a metal such as iron or calcium, or a salt-type chemical such as phosphate, which the body needs in small quantities in food to stay healthy.

Mitosis Part of normal cell division, when the chromosomes have been copied and one full double-set moves into each resulting cell.

Motor nerve A nerve that carries messages from the brain to a muscle, telling it when to contract, or to a gland, telling it when to release its content.

Mucus Thickish, sticky, slimy substance made by many body parts, often for protection and lubrication, such as inside the nose and within the stomach.

Myo- To do with muscles, such as myocardium, or heart muscle.

Nephron Microscopic filtering unit in the kidney for cleaning the blood.

Neuron A nerve cell, the basic unit of the nervous system.

Olfactory To do with the nose and smell.

Optic To do with the eye, especially the optic nerve carrying messages from the eye to the brain.

Papillae Small lumps, bumps or 'pimples' on a body part such as the tongue.

Peripheral nerves The bodywide network of nerves, excluding the central nervous system of brain and spinal cord.

Peristalsis Wave-like contractions of muscles in the wall of a body tube, such as the small intestine, ureter (from kidney to bladder) or oviduct (from ovary to womb).

Pulmonary To do with the lungs.

Renal To do with the kidneys.

Sebum Natural waxy-oily substance made in sebaceous glands associated with hair follicles that keeps skin supple and fairly waterproof.

Sensory nerve A nerve that carries messages to the brain from a sense organ or part, such as the eye, the ear, the tiny stretch sensors in muscles and joints, and the blood pressure sensors in main arteries.

Skeletal To do with the skeleton, the 206 bones that form the body's supporting inner framework.

System In the body, a set of major parts or organs that all work together to fulfil one main task, such as the respiratory system, which

transfers oxygen from the air around to the blood.

Tendon The string, fibrous, rope-like end of a muscle, where it tapers and joins to a bone.

Thoracic To do with the chest, which is also called the thorax.

Thrombosis The process of blood going lumpy to form a clot, which is also known as a thrombus.

Tissue A group of very similar cells all doing the same job, such as muscle tissue, adipose or fat tissue, epithelial (covering or lining) tissue, connective tissue (joining and filling in gaps between other parts).

Valve A flap, pocket or similar part which allows a substance to pass one way but not the other.

Vein A blood vessel (tube) which conveys blood towards the heart.

Vertebra A single bone of the row of bones called the backbone, spine or vertebral column.

Villi Tiny finger-like projects from the microscopic cells in various body parts, including the inner lining of the small intestine.

Visceral To do with the main parts or organs inside the abdomen (the lower part of the main body or torso), mainly the stomach and intestines, kidneys and bladder, and in females, reproductive parts.

Vitamin Substance needed in fairly small amounts in food for the body to work well and stay healthy.

INDEX

INDEX

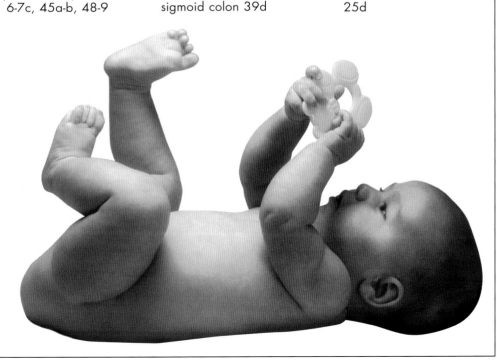

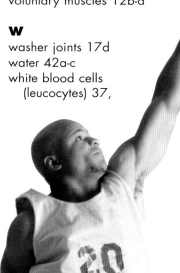